Pittsburgh's Vintage Firemen
1790 - 1915

Howard V. Worley, Jr.

Acknowledgements

One of the many pleasures of book writing is meeting people who, in one way or another, eventually become contributors to the finished work. The search for historical information is a winding trail of human encounters that connect from many different directions. As in a chain reaction, one personal contact leads to another and then another, creating a directory of information and moral support that helps carry the project to completion.

This book is no different. Even though we are blessed with modern communication tools like the personal computer, facsimile machine and answering machine, the common denominator remains the same; it is the people who bring them to life. In this case, those interested in this book took time from busy schedules to share, assist, encourage, and care about the subject matter.

Acknowledging their participation in this project, I offer my sincere thanks to: friend Richard Linder, who has loved the old fire city department since boyhood and shares my passion for Pittsburgh history; Mark Johnson, president of the national Daguerrian Society, and a student of the Great Pittsburgh Fire; The Western Reserve Historical Society of Cleveland, Ohio; Audrey Iacone and other staff members of the Pennsylvania Department, Carnegie Library of Pittsburgh, who were never too busy to answer questions or point me in the direction of valuable information; Bob Hoover, book editor of the Pittsburgh Post-Gazette, for answering my pleas for space in his column; Nancy Perlman, director of research, at the Baltimore, Maryland Museum of Industry, who generously shared her time and sources with me.

Stephen Heaver, Jr., curator, along with Debbie Brown of the Fire Museum of Maryland, for digging through their files to uncover Pittsburgh-related information; Elizabeth Lessard, historian, and the Manchester, New Hampshire, Historical Association, for sharing their archives with me; Janet Kimmerly, ex-executive editor, *Fire House Magazine*, for providing many personal contacts; The Library and Archives Division of the Historical Society of Western Pennsylvania and Librarian Cristina Lagnese; Eva Slezak, of the Pratt Library of Baltimore, Maryland; fire buff Dick Adelman, who took the time to go through his photographic collection looking for Pittsburgh fire equipment.

Pittsburgh Fire Fighters Local Union No. 1 and its newsletter editor, Fred Childs; Ed Ross, for generously sharing his detailed knowledge about city fire stations and equipment; Robert Lewis, for making available his fire negative collection for this book; Pittsburgh History & Landmarks Foundation, in particular Albert Tannler and Walter Kidney, who permitted me to use their resource library; Jim Evans of Hill's Photo Studios who once again it was my pleasure to work with bringing to life many images of the past; Federal Signal Corporation and its general manager of innovation management, Jerry Williams.

The family of the late Pittsburgh firefighter John L. Dowling; retired Pittsburgh Fire Captain Robert McIntyre; publisher Paul K. Withers for his invaluable assistance and expertise; To all of those persons who helped or contributed in any way, those who I may have inadvertently overlooked or forgotten, a special thanks; and finally, love and thanks to my wife Dolly for her inspiration, patience, and help. ☆

Cover art, **By Helen Wilkerson, courtesy of Federal Signal Corp.**
Title page art, **From the 1873 Annual Report, Pittsburgh Fire Department.** (Collection of Richard L. Linder)

In loving memory
Harriet Marie Worley
1906 - 1989

First Edition, First Printing, May 1997
International Standard Book Number: 0-9658620-0-3
Library of Congress Card Number: 97-093810

Contents copyright © 1997 by Howard V. Worley, Jr.

All rights reserved. No part of this book may be reproduced in any manner without
written permission except in the case of brief quotations embodied in critical reviews.

Marketing and distribution of this book by:
HowDy Productions, P.O. Box 445, Saxonburg, PA 16056-0445

Introduction

A little more than 200 years have passed since the settlement at the juncture of Western Pennsylvania's three major rivers became an incorporated borough. Pittsburgh has come a long way from being a wilderness frontier town to the modern, sophisticated city of today. It survived and prospered because of its most important resource, its people- more specifically, because of the sacrifices and devotion of its citizens to the place they call home.

Over time, several groups of Pittsburghers have made their life's work the safety and well-being of all residents. Among these, one occupation can trace its roots to the beginnings of the early village itself. The Pittsburgh fireman was, is, and undoubtedly will always be, a cornerstone of community service.

From a handful of part-time, untrained volunteers using crude equipment to today's paid, full-time, highly trained personnel aided by modern technology, the fireman's basic philosophy is little changed. To serve their community, save lives and protect property is the common thread that has tied all firefighters together through the centuries.

In spite of all the scientific advances, many aspects of fighting fires remain the same. Danger is ever present, whether at the fire scene or traveling to or from an alarm. The unknown, a worrisome factor at every fire, can cause untold anxiety and rub nerves raw. And the job itself is physically demanding and mentally taxing, no matter what the season. Consider the risks, not only to one's person, but the psychological toll extracted, especially from family and loved ones. So why did they do it? Why do they continue to do it?

The answer lies deep in the personal and selfless devotion to a job which puts the interests of others first. Ask yourself, is it the pay, the glory, the title of Fireman? I think not. Being a fire fighter, man or woman, requires the desire to serve in this most dangerous occupation and the dedication to go out day after day, despite the weather or other circumstances, and do the job.

It is important to recognize and remember the deeds of all Pittsburgh firemen from the age now passed on to the present. Firefighting is as old as the organized town itself, created by unselfish men who had the foresight, willingness and determination to place themselves at the disposal of all citizens in their time of need. They were the first, the standard bearers that have, down through the years, put their lives on the line each time the alarm bell sounded. To them, Pittsburgh owes a deep debt of gratitude.

Unfolding on the following pages is the story of Pittsburgh's vintage firemen, intertwined with the history of the Iron City and surroundings. One is synonymous with the other, an integral part of the life and times of the community and one of the reasons it exists today and will continue to do so tomorrow. So for all firemen of 1827 or 1927, before, in between, or after, this story is a part of your legacy, one in which the last chapter will, hopefully, never be written. ☆

> **"Good men's names are here recorded who now sleep in their graves, no more to wake till the last great fire bell shall ring the peal summoning all to God's tribunal. May they all be ready as they always were at the call of the old company bell."**
> (from the Allegheny Volunteer Company minute book)

Table of Contents

Chapter 1	Bucket Brigades and Citizen Volunteers 1790-1839	5
Chapter 2	The Crucible 1840-1869	23
Chapter 3	Getting Up Steam 1870-1899	45
Chapter 4	New Century, New Technology 1900-1915	69
Appexdix 1	Equipment	92
Bibliography		95
About the Author		96

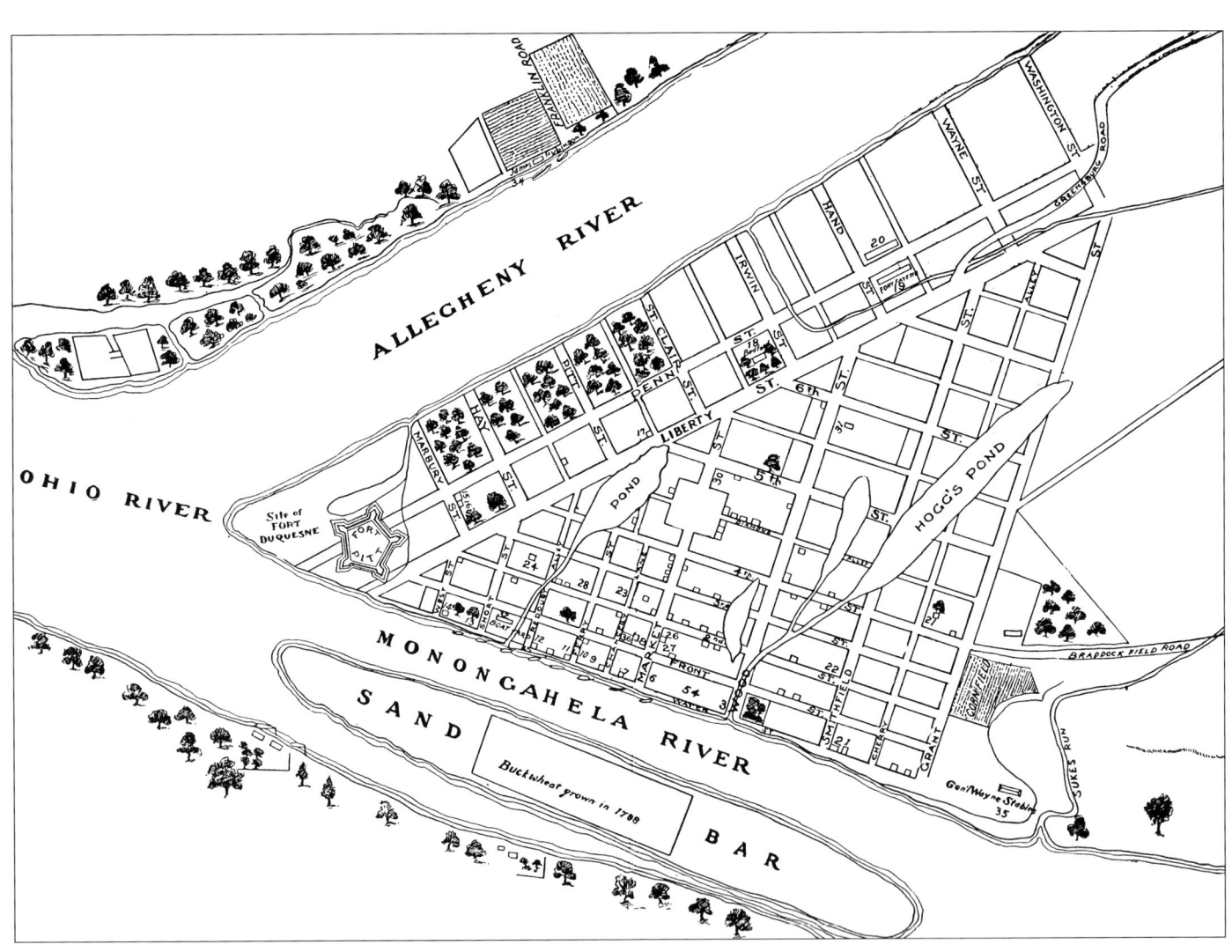

Chapter 1

Bucket Brigades and Citizen Volunteers 1790-1839

With pride, and devotion to duty, the citizens of old Pittsburgh unite to protect their town.

In the last decade of the 18th century, the western Pennsylvania settlement in and around old Fort Duquesne and Fort Pitt was a prosperous and growing town. "Pittsbourgh" in the early 1790s was the burgeoning center of commerce and population in the known territory to the west of the Allegheny mountain range.

Expansion and development on the land was a direct result of the steady influx of new "Monongahela Country" residents and the creation of new industry. In addition to its white settlers, the rustic village also included African slaves and Indians. Edge-of-town crept out from the confluence of the Monongahela and Allegheny rivers to the natural barrier formed by Hogg's Pond at the base of Grant's Hill; this tract of land was originally part of early Pitt Township. Commercial enterprise spanned, for the most part, the plateau between the river banks and was limited to the area from Wood Street westward to the revetments at Fort Pitt. Market Street was the popular focal point of most town activity. It was in this locale that several dozen merchants pursued their various trades, including coopers, weavers, blacksmiths, boat builders, clockmakers, potters, and shoemakers.

Across the Allegheny waterway, the wooded flatlands of the north shore had already been colonized by one James Boggs as early as 1760. Located close to his property by this date was the Franklin Road river ferry that transported passengers and freight to St. Clair Street in Pittsburgh. Downstream, and separated from the shore by a narrow backwater, was an island known as Kilbuck or Smoky Island.

South of the Monongahela River shore lay a somewhat level, but narrow ribbon of land. Sparsely populated, it ran from the mouth of the Saw Mill Run tributary east to a spot opposite Suke's Run on the opposite north shore. The remaining border was created naturally by the craggy face of Coal Hill, some 465 feet high. Out in mid-stream, a sand bar projected above the water line dividing the river flow into two channels. Local farmers found much success in planting buckwheat crops there, the soil being a porous consistency and blessed with the naturally high water table.

Pittsburgh town was a raw, untamed frontier community constantly undergoing change and improvement. Counted among the local structures were approximately 200 homes, 40 commercial establishments, and two churches dotting a triangular landscape. Generally, the land was laid out in 60- by 240- foot lots with east-west streets paralleling the rivers and north-south streets arranged perpendicular to the river's shore. Most of the older buildings were constructed of logs, while the newer buildings, except for a few built of brick, used locally cut mill lumber for their frames and siding. By this date, oiled paper windows were giving way to glass panes manufactured in nearby factories.

City streets were dirty, narrow, and bumpy; they were compacted mud in good weather, but an almost impassable quagmire of deep ruts, animal dung, and ponding water during rainy days. Horse-and-wagon teams clogged the avenues, passing by the faces of some buildings almost on the doorstep. Gunfire was heard frequently in the streets and back alleys of the town.

Adding to the melee were packs of roaming dogs and various forms of unhitched livestock, which ran loose through streets and yards and were known to enter any dwelling with an open door. Animal noises were heard far into the night, giving eerie life to the darkened town, illuminated here and there by an occasional oil post lamp. A blanket of thick smoke and gray haze from many coal and wood stoves hung low over the landscape day and night. This prompted some visitors to give the burgh the unflattering title of "Smoke

opposite page, **A map showing the village of Pittsburgh and environs in the year 1795. Although some streets were renamed, their original layout has changed little in 200 years.** (Carnegie Library of Pittsburgh)

This early steel engraving shows the village of "Pittsbourgh" as it looked in the 1790s. Rivers Monongahela, right, and Allegheny, left, join to form the Ohio. Dominating the background is the elevated plateau of Grant's Hill, a natural land feature that still exists. (Collection of Richard L. Linder)

Town." A layer of fine soot covered everything, giving the settlement a dingy look.

On the river, tall-stacked wooden sidewheel paddleboats churned the water, congregating to discharge passengers and cargo at the crowded Monongahela levee that ran along Water Street. Regular packet boat service between Pittsburgh and Cincinnati, was ready to begin on a regular basis. Keel boats, freight rafts, and coal barges darted to and fro, in and out, to pick up and deliver their goods at local docks, chief among them the wide landing at the foot of Wood Street.

Progress was slow but steady. Pittsburgh was a city in transition from old to new, from small to large. A regular newspaper called the *Gazette*, was first printed on July 29, 1786. Scheduled mail service, which began in 1788, enabled residents to participate in state and federal government machinery while communicating with other parts of the territory for commercial and social benefit.

Other municipal refinements evolved, due mostly to the needs of a growing population. With growth came the inevitable problems associated with it. Domestic water supplied mostly from rivers, hand-dug wells, and natural springs were being compromised by the sanitation overload from an ever-increasing human density. Town fathers were hard at work trying to develop a central, safe source of water. Public security and order were maintained using a system of roving constables, the forerunners of the later-day police patrol. Community safety was a rising concern in daily Pittsburgh life, particularly one unpredictable and constant menace, that of fire.

Both friend and foe, the ever-fickle flame was a threat, a constant worry to most local citizens and merchants. These concerns were amplified by congested narrow streets and the growing number of wood-framed, wood-roofed structures being built in the already crowded town plan. Adding to this danger was the careless and casual attitude of some people toward the use of the open flame.

Fire seemed to break out at will, anytime, anywhere, particularly at night, and without warning. If the victim was one of the frail wooden stick structures, the involvement was immediate and the damage severe. Many a man suffered permanent injury or death in a futile attempt to extinguish a fire at their homes or places of business. The common denominators that brought fire potential to all structures were fireplaces, oil lamps, and the lowly cooking stove, of which most all habitable buildings had at least one. Many shops and factories also used fire in the course of their daily work. Not to be overlooked was the occasional case of arson, whether it was an act of revenge or an attempt to collect property insurance.

Fires that erupted in the business district or where buildings were constructed close to each other in detached fashion were the hardest to extinguish, and nearly always resulted in multiple damages. Owners or proprietors had little hope of putting these fires out, for by the time they were discovered, the situation was already hopelessly out of control. Efforts were usually directed at confining the fire, trying to keep it from spreading to adjacent properties.

In addition to the swiftness in the spread of the blaze, limited means of fighting the flame and a frequent shortage of water combined to hamper efforts to save an engulfed building. Drawing water from its largest source, the river, proved to be unreliable due to low water levels during the summer and frozen surfaces during the cold months. Wells, while more reliable, lacked the

> The common denominators that brought fire potential to all structures were fireplaces, oil lamps, and the lowly cooking stove, of which most all habitable buildings had at least one.

capacity to sustain a flow of water to fight all but the smallest fires. The most critical shortage of all, however, was the lack of available manpower trained in the basic principles of fire fighting. It was the combination of these elements that led the city administration to initiate the first plans to protect life, limb and property when the fire call was sounded.

At the tolling of the firebell from the Presbyterian Church tower, townsfolk men, women, and children headed for the rising column of smoke and formed a double line. Starting at the water's source, they passed filled three-gallon buckets, hand by hand, man to man, along to the burning structure and then emptied them onto the flames. Occasionally, a ladder was raised in an attempt to put water on the upper portion of a burning building, or to try to beat out burning roof shingles. Empty buckets were then passed back for refilling via the second human line, usually made up of women and children. The efforts were herculean, the results were largely unsuccessful, and this method only served to slow the spread of flames.

By night, the setting became more dangerous and confused as visibility was reduced due to inadequate light and rolling layers of thick, choking smoke. Fire-fighting efficiency was greatly hampered as communication through the fire line was broken by citizens who deserted their posts without being properly relieved. It was a silhouetted scene of frantic animation and frenzied clamor played out against a background of roaring flames, searing heat, and feverish activity.

Afterward, the site of the fire left nothing but an ugly charred and blackened scar on the earth where once a home or business stood. Weary Pittsburghers headed home to rest and recuperate from the ordeal, hoping that another alarm would not be sounded. Many times, before they reached their doorstep, the fire bell began to toll again. These episodes were the supreme test of fire fighting's basic machinery, its human engine. In all types of weather, at any hour of the day and night, near exhaustion, with eyes burning and lungs saturated with smoke, loyal citizens staggered back to their posts. They were driven by the desire to help their fellow townsfolk and with the underlying hope that if it was their property ablaze, others would rally to help them.

As the town grew in size and population, so did the frequency and size of its fires. Clearly, a better, more efficient system was needed. From initial alarm to arrival at the scene till the final ember was out, the time element had to be shortened. Needed was a way to provide more manpower, skilled in the art of firefighting, and to deliver and concentrate more water on the fire. It was a task easier said than done. The basic ingredient in the system was the ordinary citizen, who had many other daily responsibilities. Existing apparatus was sparse, consisting of leaky oaken buckets, brooms, fire hooks for pulling down walls, and wooden ladders.

Beginning in 1792, events occurred that would improve daily life for everyone who lived in the village where the three rivers met. On September 12th of that year, money was solicited by citizen subscription to buy equipment for a town firefighting department. The handwritten receipt read: "Received of James Horner, Esq., one hundred pounds Penna. currency, to be applied in paying for a fire engine. (signed) John Wilkins, Jr., Pittsburgh, Twelfth September 1792."

On April 22, 1793, Pittsburgh officially became a chartered borough of the Pennsylvania Commonwealth. Newly elected burgesses moved quickly to enact public ordinances, with an eye to improving conditions for the some 700 residents. Many laws, like the prohibitions against firearms discharge, horse galloping, and roaming livestock, became points of humor and were largely disregarded. Several practical ideas were advanced by the town fathers which would be of benefit to every citizen. The most important of these was an attempt to organize a local fire fighting brigade.

Eagle

By mid-1793, a group of men who regularly lined up for fire bucket duty formed a social club named the Pittsburgh Fire Company. It was one of several fraternal organizations in the town at that time, among them the Freemasons and the Working Mans Mechanical Society. The stated purpose of the fire company lodge was to expand and improve the art of volunteer firefighting through social and fraternal means. Meetings were held in a rented frame building on Fourth Street between Ferry Street and Liberty Street.

City officials publicly and privately encouraged the activities of the Pittsburgh Fire Company which, in turn, raised its level of firefighting work to a more organized and dependable level. Other volunteer groups regularly appeared on the fire line, but the Pittsburgh Company was far and away the leader.

Since its inception as a social club, the volunteers had solicited funds from city businesses and people of financial means, with a goal of purchasing equipment for the fledgling company. During 1794, with sufficient monies in hand, the company placed an order with a Philadelphia concern for the manufacture and delivery of one firefighting engine. The unit was delivered to Pittsburgh in September 1794 by oxen-powered Conestoga-type wagons over the newly opened Old State Road. It was a landmark day for the town and its citizens.

Excitement and anticipation reigned as members of the Pittsburgh Fire Company started to uncrate the newfangled wonder. The unpacking was anti-climactic; there was more crating lumber

The engine was a large two-compartment wooden rectangular box, with a two-foot-deep copper-lined water trough and a centered single-action dual-chambered pump that was operated by a handle arm at each end.

William Eichbaum, engineer of the Eagle Volunteer Fire Company and Pittsburgh's first fire chief. He was also the town's postmaster and operated a printing business. (Carnegie Library of Pittsburgh)

than machine. The engine was a large two-compartment wooden rectangular box, with a two-foot-deep copper-lined water trough and a centered single-action dual-chambered pump that was operated by a handle arm at each end.

Assembly of the engine components was performed by company engineer John Johnston, ably assisted by his seconds in command, Robert Magee and Jeremiah Baker. Others who assisted with the physical tasks or offered verbal encouragement were lodge directors James Clow, William Watson, John Hanna, and James Gray. When completed, the "masheen," as it was called, sported a beautiful dark green finish coat of enamel set off by a gold stripe. It was christened "Eagle."

Light enough to be lifted by several of the volunteers, it was usually carried to the scene of a fire. There, the bucket brigade filled the two water chambers and then one husky male after another took turns working the two-arm tandem pump levers up and down. From the pump, a length of leather hose capped with a brass nozzle directed the water stream on the burning structure. At full pump stroke, a water spout 100 feet long was played on the flames, a giant stride over the antiquated bucket and ladder technique.

To house the new apparatus, the Pittsburgh Company moved into a two-story building on the corner of First Street and Chancery Lane. This location was no doubt chosen because it was in the proximity of Watson's Log Tavern, owned by Company Director William Watson. He, along with fellow officials Johnston and Magee, held the keys to the place and kept watch over the property. On the first floor, the engine waited at the ready near a wide door that opened onto the avenue. Storage space was situated in the rear portion of the structure. On the second level was divided space for living quarters.

During this time, the group officially changed its name to Eagle Fire Company and shortly after expanded it to Eagle Fire Engine and Hose Company. To all citizens of Pittsburgh it was affectionately known simply as Eagle.

Membership in the company was strictly a male affair. While many prominent names were listed on the rolls, young athletic men formed the backbone of the organization. Age and gray hair may have made good managers, but youth was the main ingredient needed when the alarm bell clanged. Equipment was pulled or carried through the streets at running speeds by the firemen. At night, volunteers led the way with torches, illuminating the way along dark, narrow streets.

Among the dues-paying members were citizens from all walks of life, notably several who were politically connected. William Leckey, who would soon serve as county sheriff, was appointed Eagle's engineer. Others included merchants and businessmen of the community. Along with men of ordinary occupations, they shared the common bond of serving their fellow townsfolk in time of need. A preference was held for married men but anyone could be considered for membership if sponsored by a current, dues-paid member. A unanimous voice vote of those present was required for election.

Aside from the business of fighting fires, the men of Eagle took great pride in their organization and delighted in displaying their engine at local parades. It was an opportunity to dress up and strut down the avenues of the borough showing off their membership and equipment. Although the company enjoyed the admiration and the attention given to them, its members also had a serious side, especially concerning their beloved engine.

It was an unwritten law that no one except a member could touch or handle the apparatus. Woe betide the individual caught violating that rule as in the case of one person who decided to test that prohibition. The company captain, a large-framed Quaker man, reportedly gave a verbal warning with "Friend, thee had better remove thy hand from our engine." When given an impudent reply, the captain promptly and without further warning struck the offender, knocking him to the ground. That incident had the town wags buzzing, giving unflattering attention to the volunteers. History does not report whether the incident was repeated.

Freeholders of the Borough held a meeting at the home of John Reed on April 16, 1797, to enact a tax to purchase fire buckets. Reading in part, the resolution stated "That a tax be laid to raise as

Pittsburgh's fire engine and hose companies formed the Fireman's Association of Pittsburgh in April 1833. These are pages from the association's first constitution. (Library and Archives Division, Historical Society of Western Pennsylvania, Pittsburgh, PA)

much money as will purchase fifty fire buckets, each marked with the letters B.P. (Borough of Pittsburgh) and be numbered from one to fifty."

At a subsequent meeting called to order on January 30, 1798, the same landowners voted unanimously to purchase Eagle's engine by levying an annual tax on each citizen for the next two years. It was the first fire engine purchase made by Pittsburgh, but it didn't change anything, because the engine was entrusted back to the men of Eagle for their use in fighting town fires.

Yet another ordinance was issued on June 28, 1798, mapping boundaries for the borough's three main fire districts. Each district was commanded by a group of men who were responsible for organizing citizen water bucket details plus other help at fire events.

Company Number 1, bounded by the lower side of Market Street to Fourth, was led by John Scull as president, with Directors John Johnson, Jeremiah Barker, William Irwin, George Adams, Oliver Ormsby, and Issac Craig. Company Number 2, taking Market Street's upper side, had Adamson Tannehill as president, and Thomas Bracken, William Mason, George Robinson, Issac Gregg, Samuel McCord, and Alexander Sholl as directors. The third company was presided over by Nathaniel Irish, assisted by Directors John Irwin, William Gray, William Dunning, William McMullin, Jeremiah Sturgeon, and Robert Griffin. It included all of the streets North of Fourth.

City fathers, meanwhile, were hard at work developing a plan to increase the available water supply needed to, among other things, fight fires in the growing community. On August 9, 1802 a town meeting of local authorities and citizens gathered to discuss the ever-increasing water supply shortage. As a result of this, Chief Burgess Isaac Craig issued an ordinance that called for the drilling of four new wells, complete with pumping equipment and stone linings; each was to be bored to a depth of 47 feet. They were to be located along Market Street; one between Front (First) and Second Street, another between Second and Third, and one in the next block ending at Fourth Street. The last was positioned in front of the courthouse.

Payment for this work was handled in a democratic fashion. By vote of those present, a general public tax was approved to cover the costs of the work which was estimated at $175 per well. Funds were also allocated for those private well owners whose facilities were often used in fighting nearby fires. All of the town's wells were supplied by several natural springs, which flowed from the higher elevations, mainly those emanating from the Grant's Hill plateau. These sources provided an adequate flow of water to all wells in use at that date.

Allegheny

During the year 1802, the second of Pittsburgh's volunteer fire companies was chartered as the Allegheny Engine and Hose Company. An immediate rivalry was struck between the Allegheny and older brother Eagle. The new company boasted the names of many well-known local citizens. No doubt some of these members were ex-Eagle men who, for one personal reason or ambitious nature, decided to start a competitive company. By unanimous vote, company members selected "Semper Paratus", Latin for "Always Ready," as their motto.

Allegheny's first apparatus was Philadelphia built, using a 10 1/2-inch double-chambered design. Its size dwarfed Eagle's machine, plus it came with 100 feet of leather hose. A separate cart, delivered at the same time, transported an additional 800 feet of hose, complete with brass couplings and nozzle. A two-story building on the south side of Fifth Street, between Smithfield and Wood streets, served as both engine house and meeting quarters.

Members of the Allegheny, like their fraternal brethren, were courageous and faithful in the discharge of their duties. Fidelity to the task and zealous devotion to their town were the hallmarks of Pittsburgh's volunteer firemen. But now that there were two lodges of firefighters, the spirit of personal and professional competition surfaced in every company action.

Naturally, getting men and equipment to the fire scene first was a point of accomplishment and honor. There were other incentives, too. In some instances, insurance companies paid bonus money to the first volunteers on a fire scene. With speed imperative, all companies jumped to be first out of the engine house. One volunteer group was quoted as saying that "They threw their boots from the second story windows and then ran down to the street, getting there before the boots landed." Extinguishing the blaze or a difficult or dangerous part of it was a victory won. Even the return trip back to where the engine was stored was done on the run, an overt show of vigor, pride, and bravado. In many instances, this led to skirmishes between winner and loser, some effects of which carried over to the next alarm.

Vilgilant

Of all the Iron City's citizen firefighting companies, the best known, after Eagle, was the venerable Vigilant. It was chronicled more than any other early volunteer service. The Vigilant came to life at the northeast corner of Wood and Fourth streets in a tavern owned by William Morrow on Monday evening, June 3, 1811. Early on, it took on the nickname "Vigy," a term of affection that no doubt reflected the sympathies of its members. But its formal name proved more than appropriate for the long and loyal service rendered by the company.

Among the charter members were listed men of the borough who would, in time, serve their fellow citizens in many important ways. These included William Wilkins, a director of the Bank of Pittsburgh and later judge, congressman, senator and United States secretary of war. Magnus Murray and John Darragh would serve terms as mayor of the city, with Murray elected twice.

Also signed in the membership book was William Leckey, who was prominent in the iron forge business. Shortly, he would become sheriff of Allegheny County. He was a veteran of the volunteer system, having served as Eagle's first engineer.

Chairman of the gathering was the Rev. Robert Patterson, pastor of the Presbyterian Church. He read the resolution to organize to those present and asked for a voice vote. The rafters and fixtures of the place reverberated with a firm cry of "yea." That done, nominations for officers were opened and a show of hands elected William Wilkins as president; John Darragh as secretary; and John Thaw as treasurer. The Vigilant was born.

The very first order of business was forming a committee of members to raise funds to purchase an engine. By subscription, $750 was collected by the winter of year 1812. A contract was entered into with the Patrick Lyon Company of Philadelphia, the most prominent equipment maker of the day, to purchase a firefighting engine. Specifications called for a "hydraulic engine of the third magnitude with all the improvements lately made."

Consigned in care of company member James Morrison, a director of the Bank of Pittsburgh, Vigy's engine left the factory in a freight wagon on September 10, 1812. Anticipated delivery time was 25 days, a promise bettered by several days when the shipment arrived in Pittsburgh on September 28, 1812. The bill of lading listed a freight rate of $98 for the 350-plus-mile wagon journey.

Company members quickly unloaded, uncrated and set up the machine. It was painted a glossy black and weighed in at 4,000 pounds, a challenge for its small diameter-wheels and direct contact brake bar. In a short time, a hydraulic test was conducted which propelled a one-inch column of water to a height of 150 feet. Finally, Vigilant was ready for duty.

Storage space for Vigy's engine was located in a frame structure on Third Street near Market. Company members continued to hold their meetings at Morrow's Tavern, where a charge account was maintained for an occasional bottle of whiskey, no doubt to oil the machinery of official business. On one occasion the Rev. Robertson's hat turned up missing and he went home without it. For some time afterward he was royally razzed about losing the hat during a lively meeting of the

---

Of all the Iron City's citizen firefighting companies, the best known, after Eagle, was the venerable Vigilant.

Vigilant brotherhood at Morrow's. Whatever their parliamentary methods were, company records indicated all member dues paid up, all fines settled and no debts owing; an enviable record to be sure.

Even though the town boasted three independent volunteer firefighting companies, most of the calls they answered were of the single-structure type. The most common cause was the errant cookstove ember or a rubbish fire that was carelessly tended. On occasion there were fires that taxed the resources of the combined companies efforts. One of these memorable events occurred on July 23, 1812, when some dwellings were reduced to a charred mass of blackened timbers. As extensive as the damage was, it could have been much worse. The fire scene was Market Street between Water and Front Streets, one of the older sections of the community. The row of structures had many rooming houses and multiple-tenant buildings. Its dense population increased the chance of human tragedy. Only through the combined work of Eagle, Allegheny, and Vigilant, and the factor of no appreciable wind blowing, was the blaze contained and extinguished.

As time passed, at least one fire brigade saw a changing of the guard in its management. Eagle's engineer, William Leckey stepped down and was replaced by William Eichbaum, well known locally for his wire and glass manufactory. For him, it was the beginning of an auspicious career dedicated to protecting the citizens of Pittsburgh from the perils of fire. Other companies were entering their second generation of members as well. Many sons of charter members followed their fathers into service, a long standing tradition practiced to the present day. As a result, family bonds developed and strengthened, along with the fraternal and social ties.

Pride in both organization and purpose did much to sustain the fire fighter in these early days of volunteer company life. Each group had its own constitutional rules that established guidelines for each member's conduct. One such rule set a fine of 12 1/2 cents for a man missing a scheduled company meeting. A missed call to duty might result in the levy being multiplied or result in expulsion from the ranks. Such a drastic remedy was used only in rare instances, and only for flagrant offenders. For the majority of the rank and file, duty was next after family and church. Many members openly displayed their lodge loyalties, like the men of Eagle who wore white ribbons lettered with "Eagle Fire Company" on their Sunday best as well as to company functions and parades.

Neptune

The last of the pre-municipal volunteer fire organizations formed was the Neptune Fire Company. It was chartered in 1815 and had its address of record located beyond the town limits, east of Hogg's Pond. Neptune held a couple of other distinctions as well. First, its fire engine was the first Pittsburgh built affair with double 9 1/2-inch chambers. Squire John Sampson designed and built the mechanical portion while John Arthurs constructed the chassis and running gear. Second, despite its official residence being outside the town, Neptune had its engine house located within the village boundaries at the corner of Sixth Avenue and Wood Street on property owned by the First Presbyterian Church.

Neptune was characterized by many locals as a "bucket company," meaning that the firemen carried water to the fire scene even before the engine arrived. A favorite location for drawing water was Hogg's Pond and its southern tributary to the Monongahela River. Even after Neptune's engine was fully operational, separate bucket lines were maintained to bring water to the fire.

On March 8, 1816, the 22-year-old borough of Pittsburgh was officially incorporated as a city. Simultaneous with receiving the municipal charter, the newly formed city was authorized by the Commonwealth to enact an ordinance requiring all building owners and tenants to furnish and

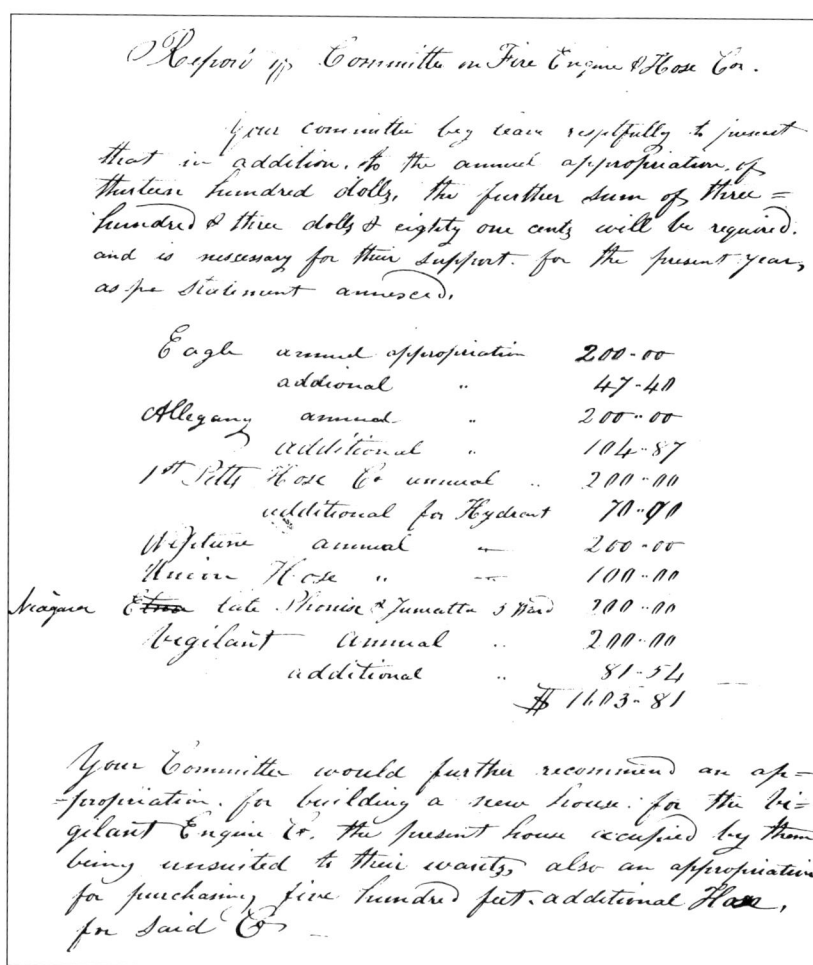

This June 11, 1838, report to the Committee on Fire Engine and Hose Companies requested operating funds for each company plus an appropriation to build a new engine house for the Vigilant Company. (Library and Archives Division, Historical Society of Western Pennsylvania, Pittsburgh, PA)

> Neptune was characterized by many locals as a "bucket company."

Citizen volunteer Timothy Tucker used this, his personalized leather water bucket, when fighting Pittsburgh fires in the mid-1830s. (Collection of Richard L. Linder)

Although Marina Betts never made the social register or participated in the finer aspects of Pittsburgh's women elite, it was said that she never missed a fire call.

maintain buckets for use in fighting fires. Councils of the town wasted little time in drafting and passing an appropriate piece of legislation. This type of citizen bucket system was first used by the City of Philadelphia before 1800.

In its simplest form, each person living in a dwelling was required to have a minimum of two buckets each, ready for municipal fire duty. To avoid the damage common to wooden buckets, these containers were actually large leather pouches or bags, similar to those used for animal feeding. They were crafted by local shoemakers. Law required that each person stencil his or her name on each bucket to ensure their proper return. This practice also discouraged citizens from exchanging their broken and leaky containers for those in better condition.

Pittsburgh was now a community of just under 1,200 homes. Commercial progress was growing steadily along with several banking institutions and churches, which erected structures in the town plan. New industry included steamboat construction, which launched the first of its kind, the *New Orleans*, in 1811.

Street lighting, now fueled by whale oil, was expanded in scope from the Market House business district outward to the residential community. On some days, the lamplighter let the lamps burn around the clock. The volume of coal smoke increased with each dwelling built, and quadrupled when new manufacturing mills commenced operations. When combined with fog, which usually hovered over the rivers on days without wind, it created a thick blanket that townsfolk called smog, a borrowing from the words smoke and fog. Street lighting was left illuminated until the sun burned the mixture off or a breeze carried it aloft.

Taverns were always an important social item on any town menu and Pittsburgh was no exception. They ranked right up there with the other successful businesses of the city. Among the local watering holes were Semple's, The Greentree, and Watson's, which vied for the packet-boat patron business as they landed on the Monongahela wharf at Water Street. In town, along the paved Market Square, William Irwin's Inn and the Black Bear Tavern had their loyal followings, among them the rank and file of Pittsburgh's volunteer fire companies.

Being a fireman, with its rank and glory, and its frequent brushes with danger and occasionally death, required a means to release occupational tensions. Local pubs served as the vehicle to do this and to bring the firefighting fraternity together to relive recent alarms and to remember the department history.

Volunteer fire companies refined their fire-scene duties through trial and error. Each member was an important cog in the firefighting machine. Assigning specific, individual tasks prevented duplication of effort while putting a maximum application of manpower into each firefighting discipline. Each company had as its leader a captain or engineer. On the way to a fire, he trumpeted the alarm to warn citizens to clear a way for the onrushing fire equipment. He alone was responsible for his company's equipment, its men and their conduct and utilization while fighting a fire. At multiple alarms, where more than one company answered the call, engineers coordinated their attack plans to ensure proper coverage without overlap.

Once at the fire, company hosemen quickly unrolled their hoses from the engine location to a central coverage point on the burning structure. Meanwhile, watermen were busy filling the engine compartments and securing a constant water supply from the closest well, spring, or other source. Axmen cut down any building overhangs that might collapse. Laddermen readied their frames to assist at second story and higher levels.

It was the duty of propertymen, with their long red wooden poles, to act as wardens, to protect personal property around the fire scene from vandals and looters, lowlife that invariably showed up at almost every fire. Each volunteer was an integral part of his firefighting team, relying on their fellow brother to perform their individual tasks and thus bring the collective company effort to a successful conclusion.

Despite the zeal and super human efforts of the Pittsburgh fire volunteers, civilian helpers were still necessary. This was particularly so in keeping the engine pump supplied with water. But by the early 1820s, interest from and participation by the public at large was declining. Some felt that, with four volunteer companies, there was no legitimate need to recruit citizen help. Others just plain didn't care. To them it was the other fellows business.

But there were some stalwarts who still viewed the bucket platoon as their civic duty, their way of contributing to neighbor and community. Local history records the tale of one loyal local bucket passer, a woman named Marina Betts. She was described as a virago; as Webster put it "a man like woman, an amazon." Standing just under 6 feet tall, with very dark complexion, she had long black hair and high cheekbones. She appeared to be of French and Indian parentage. Marina resided in the not-so-glamorous address of Shinbone Alley, the term alley in this case being gratuitous.

Although Marina Betts never made the social register or participated in the finer aspects of Pittsburgh's women elite, it was said that she never missed a fire call. When the alarm sounded, Marina was one of the first to line up. She wasn't the least bit bashful about giving her male counterparts a verbal challenge to keep up with her brisk pace for passing filled water buckets. Indeed, there were few men who could surpass her quick and lively skill at 'swinging the leather.' Idling men spectators were sure to feel her sharp tongue berate them for loafing until they either moved along or joined in the work at hand.

Those who chose to stay on the sidelines and ignore her barbs were likely to be the recipient of a bucket of water dumped over them by Betts herself. At the next fire, these dandies would keep one eye on the fire and an even more watchful eye out for one of Pittsburgh's first fire women, Marina Betts. Other female volunteers served the town, but none came close to the almost 20-year record set by the woman from Shinbone Alley.

As competition escalated among the various fire companies to gain the public's favor, each group tried to establish its own individual identity and personality. In addition to the color and trim of its engine, each brigade established its own colors to set it apart from competitors. What was developed was a standard color for the garments that would identify all brothers belonging to one particular company.

In no particular order, but with much deliberation, the firemen set about the task of deciding what color and hue would serve to represent their individual companies. Initially, two groups selected the same shade, sending the whole process back through the ranks. In the final analysis, the four separate groups came to a mutual decision. Hats were universally leather, with a tall pointed cap.

Eagle men wore green, a color they adopted shortly after their incorporation. They also designed a new hat badge to replace the existing one

A document of historical note, this was the citizens' petition to the town councils to organize the Neptune Fire Engine and Hose Company. It is dated February 25, 1833. (Library and Archives Division, Historical Society of Western Pennsylvania, Pittsburgh, PA)

they were wearing. This updated version was a white canvas shield outline with an eagle painted in black. A cream color or light tan variety was selected by the men of Allegheny. Symbolizing their vigor and the foe they fought, the Vigilant troop decided on red. And appropriate to their name, blue was the color of choice for Neptune.

Each passing day brought more urgency to the need to supply more water for Pittsburgh's expanding population and industry. By 1824, the existing system of wells and the springs that fed them were heavily overtaxed. For some time now, the City Administrators pondered a solution to the question of providing additional, reliable water service. The answer was right at the doorstep, literally, for with the Allegheny and Monongahela rivers flanking the town, the topic of quantity became secondary. It all came down to a matter of distribution.

To meet the need, Council passed legislation on February 24, 1824, authorizing the city to spend $50,000 to acquire real estate and construct a municipal water works. A five-man committee was formed to oversee this project. Funds were raised through a loan from the Pittsburgh and Exchange banks.

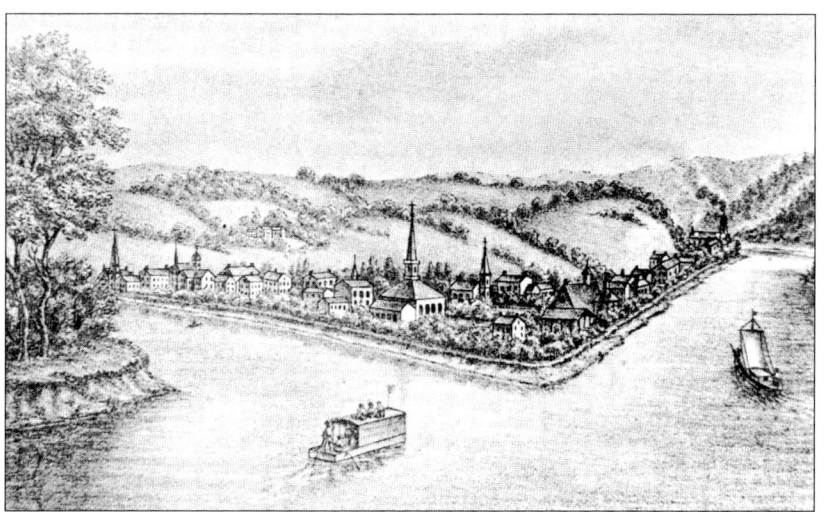

A sketch by visiting Mrs. E. C. Gibson in 1817 indicated that the town has several churches and some light industry. Period watercraft make their way along local rivers transporting people and freight. (Collection of Richard L. Linder)

Acting on City Council's behalf, Mayor John N. Snowden in February 1826 secured an additional $40,000 bank loan for a steam pumping engine and materials for the municipal water supply system. Four months later, the search was started for land on which to build the pumping station and storage reservoir. In September, the town fathers purchased a 50-by 100- foot plot of land on Cecil Alley between Penn Street (now Avenue) and the Allegheny River for $1,425. This site was designated for the pump house and its machinery. Meanwhile, a four-lot parcel bounded by Grant Street, Fifth Street (now Avenue), Cherry Alley (now Way) and Diamond Alley (later Street, now Forbes Avenue) completed the land acquisition. This Grant's Hill site, just east of Hogg's Pond, would serve as the location of the reservoir. About $3,800 was spent for both real estate purchases.

Soon after the deeds were recorded, ground was broken for the 100,000 gallon reservoir. At the same time, construction materials, pumps and some 13,000 feet of locally manufactured iron pipe were purchased. Design specifications called for the two steam driven pumps to deliver 60,000 gallons in a 12-hour period to the 100,000 gallon reservoir. This storage basin was located at an elevation of 120 feet above the pumping station. A city bond issue was floated to cover these expenses plus both of the real estate transactions. The obligation totaled $200,000, and was the largest single debt instrument incurred by the city to that date.

When this system was completed, it worked like this: water was drawn from the Allegheny River intake at the foot of Cecil Alley and elevated to the pump works, where it was transported 2,430 feet through a 15-inch pressurized pipe to the storage basin atop Grant's Hill. The completed reservoir resembled a tranquil lake with shade trees and sitting benches around the perimeter. Overseeing this new water facility was Superintendent George Evans.

From the reservoir, the main water distribution pipe ran down Fifth Street and branched off north and south at Grant, Wood, and Market streets. At cross ways and alleys, the water line was extended above ground, using a tee fitting. The pipe end was terminated above grade with a stop cock and plug in the open end. This gave firemen several locations at which to connect a water supply hose to feed their engine.

When answering alarms, volunteer fire companies raced to be first at the plug closest to the fire. Competition for municipal water was so intense that some companies sent ruffians ahead to claim the plug for them and keep others away, using any means. Many of these toughs were veteran prize fighters, which led to some citizen bestowing on them the title "plug-uglies," a label that soon became an American slang term.

With the coming of the public water distribution system, firefighting relied less on the manual bucket line. True, there were many occasions where the bucket passers were still needed, but 'passing the leathern' had passed its heyday. It was now just a matter of time until the water supply piping and fire plugs would reach into every neighborhood. The final alarm bell was about to sound for the bucket brigade faithful.

Eager to keep pace with the new water technology, the members of Eagle Company made a decision to retire their elderly engine and purchase a modern, updated machine. While the decision was made easily, the method to pay for the equipment was much more complicated. For several years following 1820, the group had fallen on less than ideal times. The fierce pride and devotion to duty that hallmarked the first volunteer company appeared to be on the wane by 1823. There was concern in the ranks about this departure from the group's avowed purpose and operational standards.

In order to restore the company to its former stature, a rededication to task and purpose was undertaken by the officers and men of Eagle. By 1825 their efforts were bearing fruit. Confidence, trust, and character were restored in the public's eye. To make further progress toward the purchase of their new engine, every Eagle man was assessed a 12 1/2 cent per month levy to help finance the venture. On this basis, company leaders approached the city councils to ask for financial support toward the new engine purchase. And the councils were in agreement, authorizing a subscription of $70 toward the $900 costs. In turn, local banks, seeing the member and municipal support, agreed to loan the remainder of the monies to complete the balance of the transaction.

There was one other important factor that helped foster the cooperative spirit between the volunteers and the town. All over Pittsburgh, politicians, businessmen and ordinary citizens came to realize that the independent fire companies were critical to personal safety and economic welfare of the community. For the money paid to each company fighting a fire, there were the intangibles received, like the devotion, pride, and professionalism that the firemen brought to the task. This was driven home in brutal fashion during an April 2, 1826, fire that erupted on Wood Street between Diamond Alley and Fifth Street. In spite of a maximum four-company response, 16 structures were reduced to smoldering ashes. It could easily have been a runaway blaze had it not been for the work of fire volunteers and their equipment.

Across the Allegheny River the village of Allegheny numbered about 1,000 citizens. Colonization did not come about quickly. Indians had frequented the land for hunting, a fact that kept most folks on the Pittsburgh shore with the river as a buffer. Uncertainty over a boundary dispute that put the territory either in Pennsylvania or Virginia further retarded settlement. In the year 1828, the town of Allegheny became an incorporated borough. Some people expected Allegheny to become the county seat of government for its namesake county, but that was not to be. Pittsburgh was named as the acting county seat, an appointment it never relinquished.

Not to be put off by the political rebuff, the smaller member of the Twin Cities forged ahead with public progress. The town council passed an ordinance on December 7, 1829 that called for the purchase of fire fighting equipment and the construction of a building to house it. By 1830, the hand pump engines Hope and Columbus, which were designed to work in tandem, took up their station in the newly built firehouse at the corner of Ohio and Federal Streets.

In contrast to the diminutive Allegheny borough, Pittsburgh was a city growing in population and expanding its residential neighborhoods and commercial sites beyond the old original landscape. To the east, the suburb of Bayardstown pushed out along the narrow plain bordered by the Allegheny River and the high ridge south of the Greensburg Road. At one time, this locale was also known as the Northern Liberties.

Along the Monongahela River's north bank there was some evidence of eastward settlement beyond Pipetown at the base of Boyd's Hill. Each day the count of new citizens entering Pittsburgh was increasing. When the Pennsylvania Main Line Canal system was connected to the city in 1829, yet another avenue for settlers was opened. All of this progress imposed additional burdens and responsibilities on the city's volunteer firefighting services.

Besides all the fires that broke out, Pittsburgh's firemen were kept in a state of readiness with many false alarms. They were inevitable and while giving the company engines exercise, they also served to sharpen the skills of the rank and file. On some of those bogus occasions, the boys took a roundabout way back to the enginehouse by way of Virgin Alley. They found it great fun to play a stream of water into the doorway of one of the more notorious hostels. The favorite target of these impromptu deluges was an address known as the Crow's Nest, a place that some citizens said needed a regular hosing down, fire or not.

Around this time, the first two fire hose supply companies started business in the City. These were the First Pittsburgh Hose Co. and the Union Hose Co. They had neither the equipment nor the skills for fire fighting work. Their purpose was to supply leather fire hose under contract to the various volunteer companies at the fire scene. One hundred feet of hose delivered more water than a 60 man bucket brigade. It was the first commercial offspring of the local firefighting service. Reportedly, employees of Bakewell's Glass Co. were pressed into service to furnish fire hose reels in addition to their regular duties at the factory.

The year 1832 was memorable for all Pitts-

An 1832 cost estimate prepared by the contractor to build Eagle's new engine house on Fourth Street. (Library and Archives Division, Historical Society of Western Pennsylvania, Pittsburgh, PA)

burgh residents, especially its firemen. In February, the second of the great river floods to visit the city raised the level of water 35 feet above normal elevation. High water was not the only concern, as rapid currents swept away everything in their path. Even the land itself was not spared, as Smoky Island, along with several resident structures, disappeared from the Allegheny channel. Volunteer fire companies were hamstrung by the swollen waters. Travel to and from fire events was a precarious undertaking because a low water route taken on the outbound trip might be flooded and impassible on the return run.

In spite of the magnitude of this event, there would be one of greater significance to the local fire brigades. It was in this year that the Fireman's Association of the City of Pittsburgh was formed. The group was organized primarily to give professional benefit and social status to the members of the charter volunteer fire companies while representing the collective interests of all city firemen and the town. A governing council of members from Eagle, Allegheny, First Pittsburgh Hose, and Union Hose Companies joined together to provide counsel and guidance to the fledgling group. For the first time, an attempt was made to bring all the major volunteer fire organizations together under one lodge banner.

During the period, the Eagle Fire Engine & Hose Company decided to search for larger quarters to house its growing membership. A search committee headed by engineer Eichbaum selected a location closer in town on Fourth Street near the Chancery Lane crossing. A local contractor's cost estimate to build a new brick building with cellar came to $1,650.00.

Important changes were in order for the Neptune company, starting with its relocation to new quarters at the corner of Sixth Street and Cherry Alley. The structure was little more than a horse barn with loft space, but it gave cover and security for company equipment and a place for the membership to gather. At this time, Neptune petitioned the city government to purchase a newer fire engine to replace the existing apparatus which had performed yeoman service for the past fifteen plus years. A contract was awarded to the Smith and Minis Company which, under the guidance of Master Builder Samuel Smith, constructed a new double-levered engine.

Upon delivery, the mechanism had difficulty pumping a steady stream of water, a point of ridicule noted by the rival fire companies. A second public test was conducted at the corner of Grant and Second streets. A festive atmosphere prevailed, punctuated here and there with cheers as first Eagle and its members arrived, followed by Vigilant and then Allegheny pulling its engine *Champion*. This time Neptune's pump delivered a stout column of water that literally dampened all the critics and sent them scurrying for dry cover.

Across the river in Allegheny Borough, the town council members voted, on July 16, 1833, to allocate $3,500 for civic improvements. Centerpiece of this legislation was a new fire engine complete with hose and a new structure which would serve as a combined firehouse and borough hall. Builder John Hamilton submitted the low bid of $900 to construct the two-story building. Detailed drawings called for two second-floor rooms, one to serve as the firemen's meeting room and the other to be used for official borough business by the town fathers. Formally dedicated as the Townhouse of Allegheny, construction plans called for a brick masonry exterior with a cupola topped roof. It was completed and ready for occupancy on January 2, of the following year. In October, the borough contracted with the Phoenix Engine and Hose Co. of Philadelphia to build another new fire engine.

Firefighting service throughout the early years of the decade followed usual patterns. Weather was always a factor to be dealt with. To overcome deep snows during winter, wooden skis were mounted under the engine's wheels to facilitate travel. False alarms were common and served only to tax manpower and equipment. Single structure fires were still the most common event and always presented the most potential for danger, due in part to their number and with the added hazard of fickle winds spreading firebrands and igniting other buildings. These fires were not to be taken lightly, lest the fire would spread to other dwellings in rapid fashion. Added into this type of duty was the occasional major fire, some with loss of life.

One such fire occurred on Friday, October 4, 1833, when an errant flame sparked an inferno at the Eagle Woolen Mills in Allegheny city. Engines *Hope* and *Columbus* put up a valiant effort but their equipment could not extinguish the flames before 13 people perished. Such was the volunteer firefighter's lot, triumphant victory one day followed by heartbreaking defeat the next. Undaunted, Pittsburgh's volunteer fire service companies carried on.

Vigilant company meanwhile was in the throes of struggling with dual problems of a general downturn in member interest and internal unrest, a situation that had been brewing for several years. Despite the efforts of company Captain William C. McCarthy, the officers and members issued a call for reorganization. On Tuesday evening, November 26, 1833, a general membership meeting was called to order at the Washington Coffee House, corner of Penn and St. Clair streets.

The expressed purpose of the gathering: to restructure the old Vigy and revitalize the membership, purpose, and mission of the lodge. Election results placed the president's gavel in the hand of James Crossan and bestowed the captain's trumpet on Dr. Jonas R. McClintock. A new constitution and set of by-laws were adopted for use on December 4th, and by years end were the operating bible for all Vigilant personnel.

The new officers set replacement of Vigy's aging equipment as a high priority. During a fire on March 16, 1834, the need became more apparent. Arriving at the Market and Third street location, Vigilant's red shirts staged their engine and began to lay hose. Center of the fire was the multi-story Bank of Pittsburgh building. Despite all efforts, the pumper was not up to the task. Shortly, the fire began to gain headway and the entire structure became involved. It was then that Captain McClintock sensed the threat of imminent wall collapse and ordered his men and their equipment to move to a safer position. It was none too soon, for they had just done so when the exterior building wall heeled over and, with a thunderous boom and earthshaking explosion, crashed to the ground. Had they not moved, there most certainly would have been loss of equipment and probably fatalities as well.

With no further justification needed, city council finally acceded to Vigilant's pleas and obligated funds for a new fire engine plus hose transport reel. The Philadelphia engine building firm of Merrick & Agnew delivered the unit by freight wagon on September 26, 1834, after receiving payment of $1,100.

It became the center of attraction when it was placed on exhibit during a parade the very next day. Hefty, it weighed in at 3,300 pounds, had large-diameter wheels, heavy-duty forged axles and hand-lever-operated brakes. By comparison, it made all other engines look small. Finish was a deep red enamel, set off with gold accent striping. Near the top crown moulding on either end was an engraved brass plate bearing the company motto: "Our Name Is Our Motto."

Dwarfed beside the new machine, Vigy's little old engine was retired, having answered her last fire call at Aaron Floyd's carpenter shop at the corner of Ross and Fourth streets on September 14, 1834. On the following June 21st, it was sold to Bakewell and Anderson for the sum of $200. They in turn later sold the pumper to the town of Wellsburg, West Virginia.

Other volunteer companies were experiencing difficulties as well. Allegheny fire brigade's officers had lost interest and fallen away from their appointed tasks of providing leadership and guidance for their men. Some even sent surrogates to represent them at company meetings, preferring not to show up themselves. It was demoralizing to the membership and a sad state of affairs. While the membership mulled over the problem, the old reliables could be counted on to carry the company business forward and to uphold the tradition of community service. These included Director, S. P. Darlington, Second Hose Director, John Herron, Fourth Hose Director, Edward Gregg, and George R. White, who capably filled the position of first engineer.

Through it all, a cooperative spirit among rival volunteer fire groups was very much in evidence. Competitors at the fire scene, and sometimes combatants after the flames were extinguished and on the way back to the engine house, fire brotherhood members were always ready to help their lodge fellows in time of need. One shining example was the Eagle company's willingness to share its quarters with both Vigilant and Allegheny during 1836. Vigy had outgrown its building due to added membership and equipment. With Allegheny company, it came down to

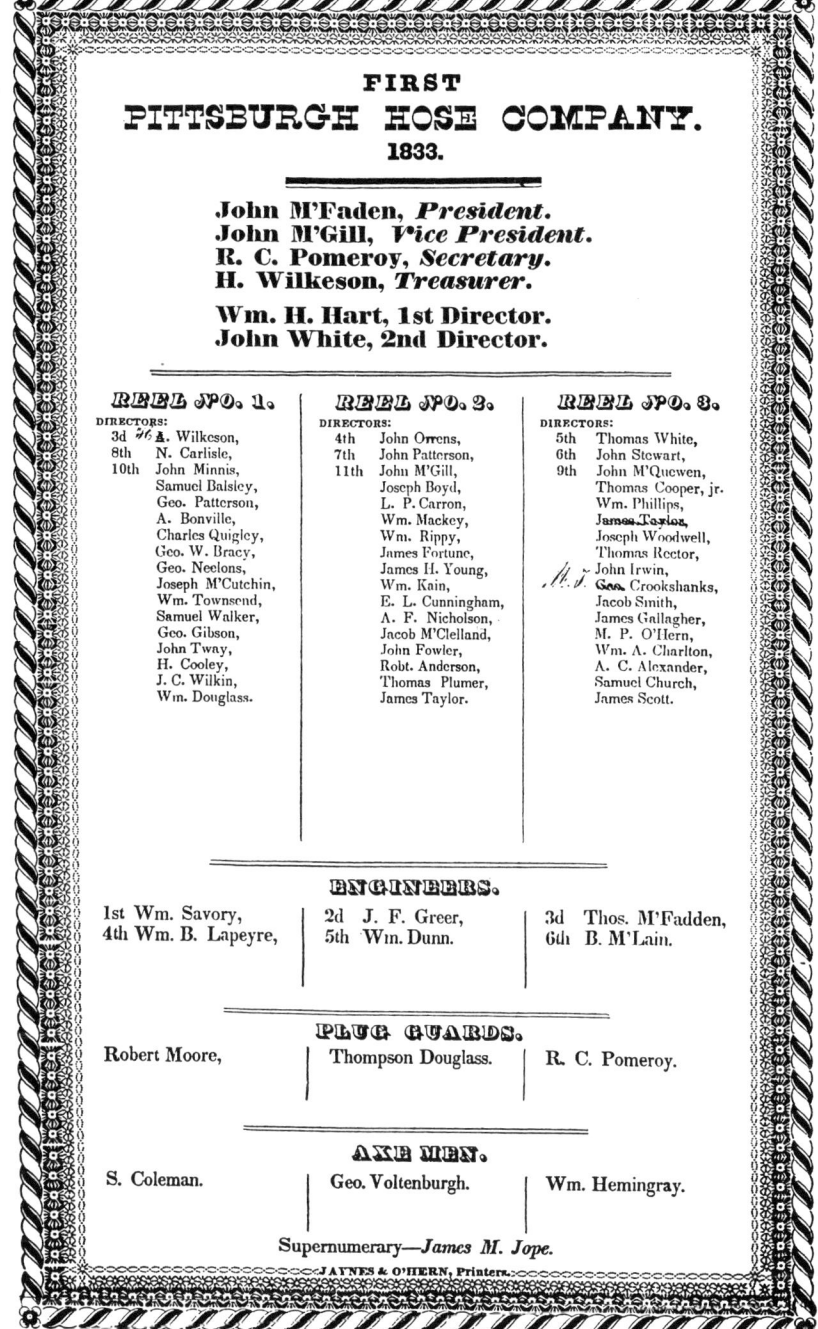

above, **Membership roll of the First Pittsburgh Hose Company, dated 1833. In all, 69 members are listed in support of the brigade's three hose reels.** (Library and Archives Division, Historical Society of Western Pennsylvania, Pittsburgh, PA)

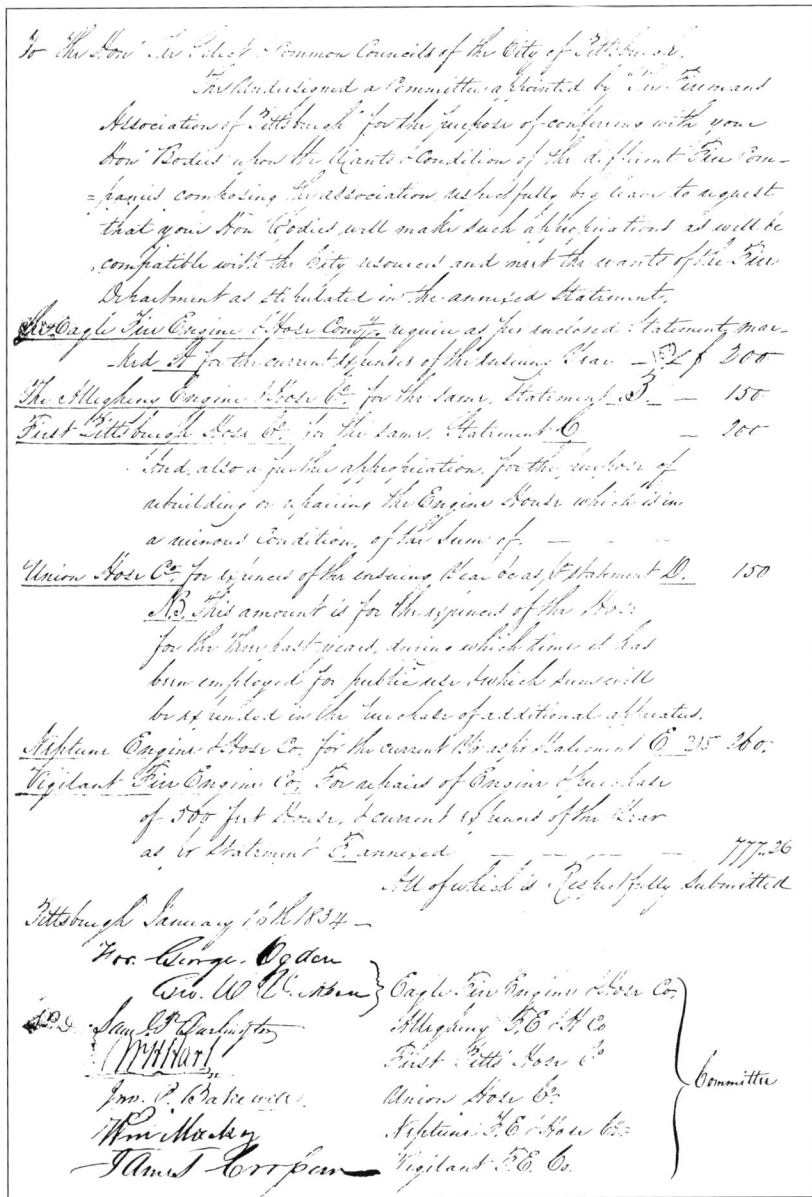

The Pittsburgh Firemens Association member companies made their needs known in this January 16, 1834, request to the select and common councils of the city. (Library and Archives Division, Historical Society of Western Pennsylvania, Pittsburgh, PA)

a point of physical safety, their building was literally falling down.

This was not to say that personal frictions were strictly inter-company affairs. Occasionally some citizen meddled in firemen's business and a rhubarb ensued. Vigilant experienced such an episode on July 14, 1836, just after extinguishing a particularly stubborn blaze. To celebrate the event, Vigy's engineer pulled up a hemp plant and placed it atop the engine. A passerby inquired as to the reason for this and received an insult for an answer. A short time later a gang of rowdies showed up to challenge the weary firemen. Fisticuffs followed resulting in the usual blue, black and purple badges of honor on both sides.

Fights and arguments between competing volunteer fire companies were regular occurrences; sometimes before fighting the fire, more likely after the fire was out. Then the boys were tired and short-fused. Alcohol was a frequent catalyst, firemen were known to carry spirits with them in case a bracer was needed. Any trivial thing might start a ruckus. On many occasions provocation was deliberate. There seemed to be particular enjoyment derived from fistfighting. Drawing blood was the manly and honorable thing to do. Whether the reasons were personal or professional, these street brawls, although viewed by some as a form of public entertainment, literally gave the city a black eye.

Company officers attempted to intervene and control such outbreaks but many times they themselves became embroiled in the fray. One company, Allegheny, attempted to take a first step towards curbing such behavior. In January 1837 they passed a resolution that asked company members, "To make as little noise as possible both at fires and false alarms, particularly when such fires occur on the Sabbath." It was one attempt to avoid trouble before it could get started.

That same year the Fireman's Association of the City of Pittsburgh met and elected W. M. Shinn as President, William Eichbaum as first engineer; John McFaden as second engineer, H. D. King as third engineer, George R. White as treasurer, and J. P. Bakewell as secretary. Membership in the association consisted of nine separate groups. Listed on the rolls were Eagle Engine & Hose Company, Allegheny Engine & Hose Company, Vigilant Engine & Hose Company, Neptune Engine & Hose Company, Union Hose Company, First Pittsburgh Hose Company and the newly formed Phoenix and Juniata Engine & Hose Company. Alleghenytown was represented by the combined Hope and Columbus Engine Companies and the Phoenix Engine Company.

The association was instrumental in standardizing the type of hat, or helmet, worn by Pittsburgh's volunteer firemen. Following the styles used in other cities, the metal crown curved upward to a pointed top while the brim flared out from a narrow point over the brow to a wide protective part covering the neck. Affixed to the cap front was each company's badge or emblem. This style of hat, modern derivatives of which are in use today, served another very important purpose. If a fireman was trapped inside a burning building, he would throw his helmet through a window to the street below, signaling to his comrades for help.

Each volunteer company elected three delegates per year to sit on the association's board of directors. Counting all companies, the association represented more than 900 firemen. Their combined equipment comprised eight pumping engines with a total combined hose length of 6,580 feet carried on 10 carriage reels.

The year 1837 brought the election of the following Neptune Engine & Hose Company officers: John McQueen, president, John Irwin, vice president, R. Porter, secretary, and T. Myers, treasurer. Captain and leader of the 80 man squad was William Edgar. During the year, the 165-member Eagle brigade listed J. D. Davis as president, G. W. Jackson as vice president, Thomas Marshall as secretary, J. B. Bell as treasurer along with John Hays, captain.

Allegheny Engine and Hose Company officers included W. M. Shinn, president, Samuel C. Hill, secretary, and George McCandless, treasurer. Captain of the 97-man company was Samuel P. Darlington. In the Northern Liberties, the Phoenix & Juniata Engine and Hose Company was led by Captain John Ralston, Lieutenant Joseph Dripps, and Engineers Levi Sovereign, William Day, Daniel Zimmerman, and William Harmagh. G. A. Martin served as company secretary.

Both of Pittsburgh's hose companies were fully operational with their officers duly elected. Union Hose Company numbered 40 members, among them Thomas Bakewell, president, J. B. Champ, secretary & treasurer, J. P. Pears, director, and J. B. Bakewell, captain. Their equipment consisted of a two-wheel hose carriage that transported 700 feet of hose.

The larger First Pittsburgh Hose Company provided 700 feet of hose on its four-wheel carriage, "*Fame*," and 500 feet of hose on the smaller two wheeled rig, "*Pilot*." John McFaden served as president and chief engineer, William H. Hart was vice president and director, R. C. Pomeroy filled the duty as secretary while H. Wilkeson held the post of treasurer. A total of 62 members belonged to this company.

By August of the year, Vigilant's membership had grown to 205 able bodied men. In the president's chair was Peter Baird. C. Plumb, W. S. Savely and S. Wardenbough filled the posts of vice president, secretary, and treasurer respectively. On the trumpet was Captain Jonas P. McClintock. This company was also in the forefront of experimentation to boost water pressure for firefighting. One attempted method was to pump water from one engine to another, starting at Liberty and Markets streets and ending at the Allegheny diamond. In theory, this was supposed to boost the line pressure and increase water volume delivery, however, it was a dismal failure.

Tests continued with two engines, one supplying water and the other playing it on the fire. This method showed more promise and also led to another of firefighting's honor-code quirks known as "washing." If the first engine pumped faster than the second engine could deliver, water overflowed the trough sides, hence the term "washing." No company wanted to suffer the embarrassment of being washed and feel the stigma of dishonor that it implied. The only way to lose the awful tag was to "wash the washer" back, and return the dreaded curse.

Vigilant did make progress in one other important area. When firemen answered a night call, they did so under the added danger of darkness. Trying to navigate the narrow streets and tight corners was tricky even in daylight. At night, with little or no light, collisions between equipment and citizens or property happened frequently.

City council had been petitioned by Vigilant on several occasions to install a lamp over the enginehouse door to give notice that a fire call was or would soon be in progress. When this met with opposition, the company enlisted the aid of responsible neighborhood lads who kept a torch ready, lighting it on alarm with it lit to lead the engine and firemen to the scene of the blaze. Running ahead of the machine, the boys shouted "Fire!, Fire!", warning everyone along the route that fire equipment would soon be coming. The torches provided just enough illumination to help the firemen steer their way to the fire.

Vigilant was one of several companies to have on its rolls that special four footed canine member close to the hearts of all its firemen, their mascot. Vigys was named Bill, an English bulldog who couldn't have been more representative of the breed or more appropriately suited to the life of a firehouse dog. Bill, faithful friend to all the red shirts, was most loyal when the fire call came. At

By 1837, Pittsburgh was engaged in the manufacture and sale of fire hose. This woodcut of the tandem fire engine and hose reel illustrates the latest design of equipment in use during the period. (Collection of Richard L. Linder)

The Penn Fire Insurance Co. of Pittsburgh placed this fire mark on all buildings that it insured. It guaranteed that volunteer firemen would be paid for extinguishing a fire on the property, as well as serving a deterrent to arson. This cast iron fire mark featured a bust of William Penn and was used during the years 1841 to 1845.
(Collection of Richard L. Linder)

the first bell clap he would take up station at the enginehouse door before many of the faithful red shirts arrived. When the men and machine departed for the fire, so did Bill, trotting right along with the group.

After a time, some of the men feared for Bill's safety, lest he be trampled in the rush to the blaze. Confinement sounded like a good idea to keep Bill safe. At the next alarm, while he was kept in a locked room, the fire bell sounded. Bill reacted instantly and exited the space, crashing through a closed window taking the broken sash with him around his neck.

Later, while he was quartered at R. G. Buford's Wood Street building on the third floor, Bill's keen hearing detected the fire signal. He used his favorite exit path, this time an open window at the attic level, and sprinted across rooftops to Third Street where he entered a building through an open window and sped down the stairs to street level. Recognizing the red shirted brigade and its engine going down the street, Bill raced after them, taking his place in stride with the running group at Wood and Fourth Streets.

In spite of his devotion to duty and company, Bill had one very annoying habit. He had a penchant for nipping at the ankles of rival company members or anyone who he thought was too close to the company machine or personnel. For his own safety, and to keep harmony in the fire brotherhood, Bill was retired to the country to live out the rest of his days.

By this point in time, age was beginning to show on the municipal water system, now almost ten years old. While operating capacity was nearing its maximum point, demand for water service continued to increase as city population and industry grew. During July 1838, the main pumping station was operating 20 hours each days. This put an average of 1.5 million gallons into the system during that period. Water service was being provided to 2,679 buildings at the time.

There were serious concerns about the Grant's Hill reservoir from a safety as well as supply standpoint. Anticipating future water demands needed to keep the city operating, the town council directed that land be purchased on which to build new reservoir facilities. Issuing loan certificates, the city purchased three separate land tracts for a total of $40,000 in August of 1838.

Largest of the parcels was situated at Elm and Prospect streets, on a portion of the O'Hara estate. Purchase price was $25,000. The other two lots included one in the Laughlin plan, for which the city paid $12,500, and a small fifth-ward lot for $2,500. Moving quickly, water works Superintendent Robert Moore advertised for contractor bids to grade the Prospect Street site. Public bidding notices were posted in November 1838, with pricing letters from contractors due in December.

Niagara

This last member of Pittsburgh's "Famous Five" volunteer fire companies, the Niagara Fire Company, was formally chartered during 1838. It was the successor to the year-old Phoenix and Juniata Engine & Hose Company which existed on the incorporation documents only. Their official address was along Liberty Street in Bayardstown. John Stewart was listed as company captain.

For whatever reasons, the Phoenix and Juniata Company disappeared about the time that the Niagara volunteers first appeared. A single-story wooden-framed building on Penn Street close to 15th Street served as their engine house. Tanner John Ralston was installed as president while Colonel Samuel McKelvey became its first captain. Niagara's motto was "Star of the West."

Company volunteers drew on the expertise of the E. & F. Faber Co. to construct a new tandem pump engine. Faber's shop was situated along Penn Street at 28th Street, close enough to allow company members to look in on the builder's progress. Once delivered and operational, Niagara men and their engine became known for being

Pittsburgh's river commerce activity was centered at the Monongahela River levee, shown here in this 1840s drawing. During the grat 1845 fire, a wall of flames swept up from the west to consume everything stored there. (Collection of Richard L. Linder)

prompt in answering alarms and performing their work in a competent manner. Future laurels would lie ahead for the Niagara Fire Company.

Meanwhile, Allegheny Company was kept busy fighting fires of a different sort. Once again, internal politics and public complaints about member conduct consumed many hours of company time. With sides taken, little progress was made at the regular membership meetings. The protracted circumstances were wearing down company management and placing a negative aspect on firefighting work.

Officers came and went. Some, like S. P. Darlington, asked to be excused from the February 5, 1838, meeting because he had to attend an auction sale. Others, like banker, James B. Murray, second lieutenant, Jared M. Brush, third engineer and future mayor; plus vice president, Robert M. Riddle, editor of Pittsburgh's Commercial Journal, attempted to restore order and stability to the internally fractured company.

As the decade came to a close, so did the first 50 years of Pittsburgh volunteer firefighting. It was an era of growth and progress, one of new technology and change for the town and its citizen firefighters. Both had come far, accomplished much. Each would depend on the other in the years to come, where challenges great and small would test their cooperative efforts. The trial by fire was yet to come. ✯

GREAT CONFLAGRATION AT PITTSBURGH PA.
APRIL 10TH 1845

Chapter 2

The Crucible
1840-1869

In spite of the great fire, the city rebuilds itself and prospers.

With the dawn of the 1840s, the Iron City's landscape showed significant change from even a few years earlier. Hogg's Pond and the lake at the intersection of Liberty Street, Fourth Street, and Jail Alley had vanished, pumped dry and filled in with earth from Grant's Hill. The canals that linked them to the Monongahela River were tamped firm to keep the river within its banks. Lesser ponds at Wood Street, between Second and Fourth streets, and near Market along First Street, were also drained and backfilled.

Additional new building lots were laid out on the reclaimed land and promptly sold to eager buyers. New homes and businesses sprang to life, filling in gaps between older buildings that had bordered the water's edge. Streets long interrupted by the visiting river were at last joined. With the horizontal expansion on the land came new structures growing vertically, from single or double floors to heights of three and four stories.

While this progress was welcomed by townsfolk in general, the city firefighters faced new and potentially more complicated fire situations. Increasing a building's height by 10 or 15 feet put it beyond the reach of existing ladders, and that much farther from the fire hose nozzle. Constructing a home or business on every lot along a street eliminated any natural fire breaks and prevented firemen from attacking a blaze from the side or rear. With progress came problems not apparent to the average citizen.

Public expansion was not limited to the city triangle between the rivers. A settlement was spreading east and west along the south bank of the Monongahela River at the base of Coal Hill. Known as South Side, the area prospered as settlers left the overcrowded city for the rural setting that homes and farms shared with light industry and numerous coal pits. On the north shore of the Allegheny River, Allegheny City, with more than 10,000 citizens, finally received its charter in 1840. One of the first orders of business was organizing a municipal fire department. Much growth was evident there, with land being developed northward toward Hog Back Hill. From all points of the compass, the Pittsburgh area was growing outward, across rivers, to the far shores, and beyond.

But progress was coming at too fast a pace. A heavy burden was being placed on municipal services, already taxed with trying to keep up with existing needs. New building construction alone was growing at a feverish rate, packing lumber and human beings into any available open spot on the land. What was not growing with the times, indeed it was almost remaining status quo, was the state of the volunteer firefighting system. Both in readiness and capability, the firemen and their equipment had not blossomed with the times. This combination of circumstances was ripe with potential for disaster, and increasing in probability with each passing day.

The metropolitan Pittsburgh area included the city itself, from the old forts at the river's point east to Bayardstown on the Allegheny, and just beyond Kensington, also known as Pipetown on the Monongahela shore. Allegheny was linked to Pittsburgh by the Hand Street Bridge (now 9th Street) built in 1839 and the first timber bridge across the Allegheny, which ran between St. Clair and Federal Streets. This span replaced the old ferry boat service which was the first public conveyance between the Iron City and the Allegheny north shore, some 1,100 feet distant. South Side's sole link to town was the covered wooden Smithfield Street bridge. In addition to citizen and commercial traffic, these bridges served to connect the area settlements so that fire equipment could travel from one location to another to lend assistance during major fires.

A total of 11 volunteer fire companies and support groups were on active service in the

opposite page, **Currier's print recorded the Pittsburgh fire of April 1845. Just to the left of the flames is the tower of the Third Presbyterian Church; to the right, burning fiercely, is the Monongahela House Hotel.** (Carnegie Library of Pittsburgh)

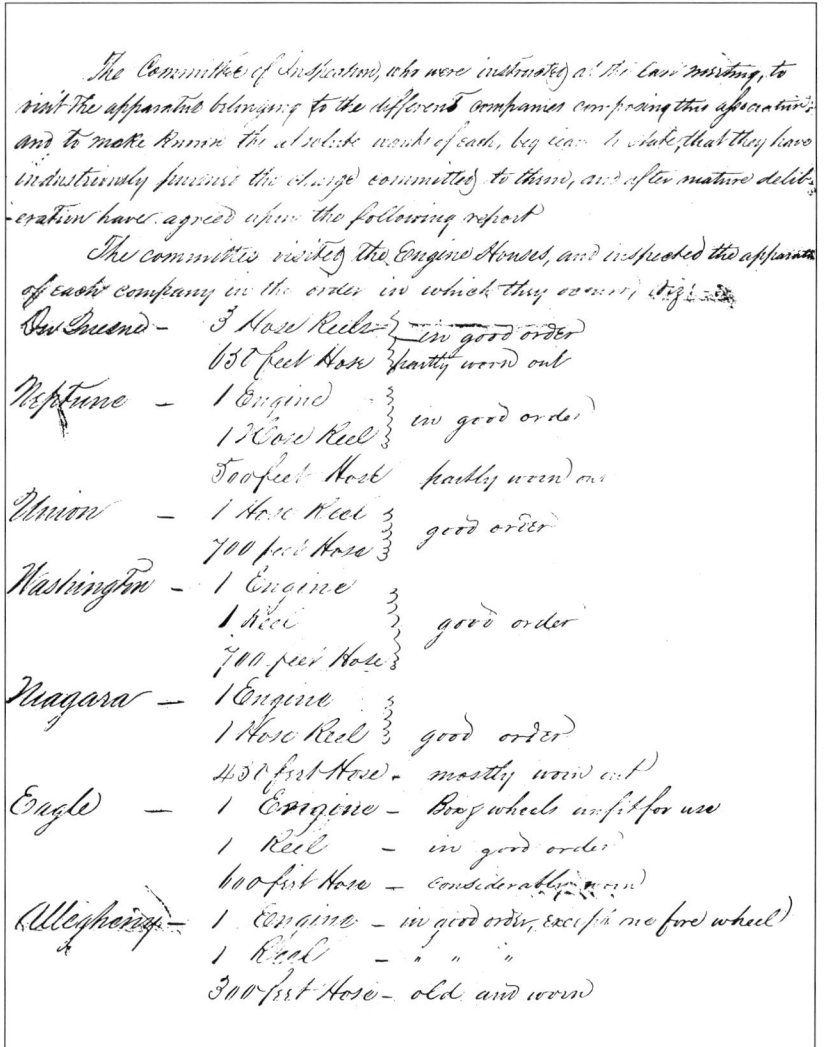

The City Committee of Inspections forwarded this August 22, 1842, report on the status of volunteer fire companies to the Common and Select Councils for action. (Library and Archives Division, Historical Society of Western Pennsylvania, Pittsburgh, PA)

combined area. Pittsburgh proper was home to Eagle, Allegheny, Vigilant, Neptune, and Niagara engine companies, assisted by hose companies Union and Pittsburgh. South Side was supported by the South Side Hydraulic Company and the recently organized engine company named Hope. Across the river in Allegheny City were two active engine companies: the Phoenix and the Hope & Columbus. The ratio was almost one engine company for each two square miles of populated area. Four volunteer companies including Washington, Uncle Sam, President, and William Penn supplied the necessary manpower.

During the early forties, Allegheny Company was in a constant state of turmoil. Several civil actions were filed against the organization to collect various debts. By the time all judgements had been satisfied, the management of the company had changed entirely. A reorganization of officers was effected in September 1842. James E. Wainwright was selected to fill the president's chair, with James B. Sawyer elected as secretary. Line officers included Alexander Richardson, captain, and William H. Whiting, hose director.

Drinking of hard liquor while on duty continued to be a complaint lodged against the tan-shirted firemen. Anxious to dispel the notions of such behavior, the Allegheny volunteers adopted, in October 1842, articles amending their charter that stated: "Members will not make use of any intoxicating drinks while acting in the capacity of fireman." Additional language that read: "When refreshments are offered at a fire, the volunteer members will go to the house containing same under the direction of company officers, who will admit members only," was also included.

Imbibing on the job was a hard-to-break habit. Volunteers drinking in the social settings like parades, banquets, and picnics was next to impossible to monitor. Things eventually got out of control. As the drinking went on, old rivalries and animosities surfaced. Mass fights pitted one volunteer company against another, rank and file against officer, rich man against poor, citizen against citizen. Next to the pleasures of drinking beer and liquor, the volunteers took great pleasure in fighting.

Typical was a Sunday tussle between Eagle and Allegheny companies. On the way back to the enginehouse, the men in green happened upon Allegheny brigade near Liberty and Market streets. One rowdy couldn't resist the chance to provoke a fight. Eagle's captain, determined to avoid trouble, quickly borrowed a pitchfork from a nearby hardware store and chased the troublemaker down the street. This situation was typical during those years when intercompany fighting reached its peak.

For Neptune, equipment concerns took first priority. A rebuilding of its tried and true Smith and Mills engine was necessary. It was worn out from years of service and needed a thorough overhaul. Modernizing included removal of the water box troughs, adding instead dual hose couplings at the pump case to accept water from hoses connected to the municipal water plugs, generally known by this date as fireplugs. On December 31, 1841, the refurbished pumper was returned to active duty.

Neptune was also making progress in attracting men to its ranks. Sixty seven new members were added to the rolls during 1842. That same year, the company organized the Neptune Library Association, which cataloged many books donated by the members. Following in 1843, 71 new volunteers took the pledge to uphold the honor of old Neptune Company and serve their city.

One of the first Pittsburgh fire-related inventions to be developed in Pittsburgh was devised by Neptune member F. S. Turbett during 1843. Considered automatic for its day, the device, when activated by hand, would ring an outside station alarm bell to alert all personnel to report for duty. It was short-lived, however, as anyone could trip the device, setting off false alarms.

While dame fortune smiled on Neptune, the

A contemporary view of Pittsburgh from the mouth of Saw Mill Run just before the fire of April 1845. (Carnegie Library of Pittsburgh)

Vigilant Volunteer Company fell on hard times. Rancor had settled deep within the organization causing a rift in the membership. A splinter faction went its own way, creating the Duquesne Volunteer Fire Company in 1842. Remnants of the old red-shirt brigade struggled on and in February 1843, with a new constitution in hand, old Vigy was ready to answer the fire call once again.

With the population growing and the settlement expanding, domestic water demands mandated construction of a new water storage and distribution system for the city of Pittsburgh. A site was chosen along Bedford Street, high on a 400-foot overlook facing the Allegheny River. This location had the local nickname "Stone Quarry Hill," due to large outcroppings of rock along the land paralleling the river. Excavation began at the reservoir in 1843 and within a year, the pumping station was delivering water from the Allegheny River intake by pipe to the hilltop facility.

Heading into the middle of the decade, Pittsburgh 's volunteer firemen continued to perform loyal dedicated service. In spite of their many trials and tribulations, the determination, resourcefulness and professional skill level of these companies continued to be the common denominator between them all.

As an example, consider the plight of Neptune. During the autumn of 1844, the fire bell sounded for the Second Presbyterian Church on Diamond Alley. Neptune's engine was out of service due to a broken axle that had been removed from the chassis and taken to the blacksmith's forge owned by Marshall and Co. Commandeering a nearby wagon, the firemen loaded their crippled engine and, stacking kegs of tobacco beneath to keep it level, headed for the fire.

But Neptune paid a high price for this inventiveness. At the fire scene, the wagon was hemmed in tight on all sides in the alley and, at the height of the blaze, a wall of searing heat overcame them and blistered off the engine's blue paint, right down to the wooden frame. Rescued in the nick of time from his perch atop the engine was Neptune's mascot. The canine was named, appropriately, Neptune, a close-cropped terrier that sported a handmade silver collar. Fortunately, the equipment was not destroyed, only singed around the top and sides.

GREAT FIRE ! ! ! ! . . . . CONFLAGRATION! ! ! ! !
(Written in the diary of city Attorney Robert F. McKnight, April 10, 1845)

In Pittsburgh, April is a time in which all four seasons can appear within the space of thirty consecutive days. The weather is fickle, unpredictable and surprising. Making the transition from winter to spring can bring bright sunshine and warm breezes, elevating the temperature into the 80s. On the other hand, record snowfalls and chilling cold have arrived on more than one April Easter Sunday. These weather patterns have always been dictated by one of nature's more formidable forces, wind. In conjunction with the local topography, the Pittsburgh winds work to create a variety of weather conditions subject to rapid and severe changes.

Springtime of the year 1845 was welcomed by the citizens of the Iron City and environs. Melting snows of winter had receded into deep ravines and sheltered valleys by the end of March. With April came warmth and a sparkling freshness to the air. Although the gray of winter was not entirely gone, bright sunlight gave life and color to the surroundings, urging people to go out of doors.

In the city, the balmy weather encouraged shopkeepers and home owners to prop open a door or raise a window sash to capture the warm breezes. Doing chores outside was a welcome change after the long confinement of winter. Here and there a saw could be heard making its way through a log. Somewhere a hammer pounded out the cadence of a carpenter's work. Even the animals were relaxed and basked in the comfort of old sol and fresh, spring air.

As April entered its second week, days of intense sunshine and warm southwesterly air currents prevailed. No rainfall had fallen for almost four weeks. Daytime temperatures hovered in the 70s. Pittsburghers looked forward to each new day and the fine, fair weather it was sure to bring. It was if the fever of spring had descended on the entire community. An air of complacency was developing among the citizenry and attitudes were becoming whimsical, factors that were fueled by the bright new season. But all was about to change in dramatic and tragic fashion.

The morning of Thursday, April 10th, dawned to fill the sky with more bright sunshine. A stiff warm breeze quickly rose, coming up the Ohio River valley from the west, raising dust swirls in the dry, parched-earth streets of the town. At the south and east corners of Ferry and Second Streets, one Mrs. Horne or Mrs. Brooks, depending on which legend is correct, was preparing for her day's work as a washwoman, a laundress. It being too hot to build a fire inside, she, instead, decided to build a small wood fire in a vacant lot shared with her home and the rear of an icehouse owned by William Diehl.

After kindling her fire and setting the water-filled iron cauldron, she promptly left to tend other chores. Around the dinner hour, the wind finally subsided. The flames from the fire, which had been warming the wash water, began to lick the dry grass leading to the base of Diehl's building, finally igniting it. Moments later, the flames ran the length of the plank wall at grade and then began to travel upward on the dry, brittle wood siding. A column of black and white smoke rose straight up, pointing a jagged gray finger into the blue April sky. Within minutes, two walls of the icehouse became involved with fire. Then the unthinkable happened. From west by south, the wind came up suddenly with increased velocity and warmer temperature.

Alerted by the unmistakable smell of wood smoke, the washer woman ran around the corner of her house, where she was met by a wall of flames. Overhead, the warm air was drawing up the cooler layer of air at ground level. The action was cyclonic. As the air came up so did the flames, smoke and embers with it. When the fire cleared the ice building parapet, the roof surface ignited and then the wind, finding no resistance, proceeded to lift up the burning building embers, distributing them on neighboring structures. Diehl's icehouse was now totally involved by a fire that was out of control. The woman's home was a burning shell of charred lumber. It was now about a quarter past noon.

Heading up Second Street, the fire rapidly consumed William Diehl's two-story frame residence and then enveloped a second and third wooden dwelling in quick succession. A wide, dense plume of smoke, spread by the blowing wind, alerted the volunteer firemen. Alarms shouting "Fire!" were quickly sounded throughout the neighborhood.

Vigilant was the first volunteer company on the fire scene. Coming on the run from their station house, they took up a position on Ferry Street between Front and Second Streets. A hose was quickly run down Front Street to the water plug at the intersection with Market Street. Cutting through a yard at the corner of Front and Ferry, the volunteers opened their valves and played the water stream on the burning structures, including the residence of Captain James May. Shortly, Union Hose Company, under James McDonald's command, drew up and connected a water supply hose to help feed Vigy's pump.

Suddenly, water pressure in the hoses began to drop, and then stopped altogether. A quick examination of the Front and Market Street plug found it bone dry. The main feed pipe from the reservoir had been emptied. In no time, the fire gained the upper hand, roaring down Second Street with a fiery vengeance. In its path, the homes of many local physicians and prominent citizens were leveled.

Meanwhile, burning embers from the three Second Street homes were blown across the narrow street, showering the Globe Cotton Factory roof with firebrands starting several independent fires. Before owner James Woods and his employees could muster their forces to fight the fire, it had taken a foothold and spread to other parts of the factory. In a short time, a roof section burned completely through and collapsed, sending hot embers down into the factory interior stacked high with yarn goods. The building was doomed and, after removing company records and a few valuables, Woods surrendered his building to the raging wall of fire.

By now, the alarm bell in the Third Presbyterian Church tower was clanging furiously, alerting the population that a fire of major proportions was in progress. When the Globe building walls finally collapsed, the burning debris fell on the adjacent home. This was the residence of the Johnston family that owned and operated

---

In no time, the fire gained the upper hand, roaring down Second Street with a fiery vengeance.

Johnston & Stockton's book store on Third Street. It caught fire immediately. Superheated by the adjacent blaze, the wooden roof ignited on impact. Even though it was a brick structure, the blistering heat of the onrushing blaze was claiming everything in its path, adding fuel to the growing firestorm. Combustibles were reduced to ashes, and even brick and stone were baked and split beyond recognition.

Due to the changing wind direction, and intense heat waves emitted from the fire center, the flames were spreading in two opposite directions. Next in line on the northwest flank of the fire, at the corner of Ferry and Third streets, was Pittsburgh's tallest structure, the Third Presbyterian Church.

Just as the fire turned onto the church property, the Eagle volunteers were pulling their engine at a furious pace along Ferry Street. Niagara Company was not far behind. At the corner of Ferry and Third, they set their engines in tactical position and began to lay hose. Quickly, a hose coupling was attached to the nearest water plug and the valve was cracked open. A disappointing trickle of tan water exited the hose nozzle and made a mud puddle in the dirt street. A full turn was laid on the valve wrench which produced a weak stream of dirty water. Pressure was low due to reduced water levels in the Grant's Hill reservoir.

Eagle Company's main efforts were directed to protecting the fire break between the crackling mass of fire that once was the Johnston residence and the church. What other water was available was played against the church's east wall by Niagara to keep flying sparks and airborne fire debris from igniting the elderly landmark. It was a touch-and-go situation for awhile, the intense heat and flames finally winning the battle and setting part of the church roof cornice on fire. Eagle's men attempted to play water on the area, but the feeble stream couldn't reach. Congregation members aided the firemen in trying to turn back the fire. Several firemen crawled out on the steep-sloped roof to beat out the flames, but were repelled by the heat and the difficulty of trying to keep from falling over the roof edge.

In desperation, Eagle's volunteers hacked off the cornice framing members with axes, letting the fiery mass crash to the ground in a giant shower of sparks. The tactic worked and shortly afterward, the wind changed direction again, pushing the flames eastward. Pittsburgh's landmark church with its 163-foot steeple was spared. Others weren't so fortunate.

Along Market Street near Front, Neptune was locked in a struggle to save the four-story Wickersham residence. When the water plug dried up, the Blues resorted to a bucket brigade. A human line was set up from Front Street to the river a short distance away. In spite of their efforts, the fire gained a major foothold, forcing

For many years, the Third Presbyterian Church was the tallest building in Pittsburgh. The landmark structure was saved from destruction in the 1845 fire by heroic work of several volunteer fire companies.
(Carnegie Library of Pittsburgh)

Neptune to retreat. In short order the home was totally engulfed in flames.

Attempting to get ahead of the blaze, both Vigilant and Neptune moved to new attack positions. Old Neptune set up its machine at Third and Market streets in an effort to contain the fire and keep it from spreading across Fourth Street. In the process, the Post Office was saved but the Bank of Pittsburgh was lost. Ordinary citizens were now standing shoulder-to-shoulder with the firemen, trying desperately to control the flaming monster. Among the helpers was eighteen-year-old Stephen Collins Foster, whose songwriting fame had yet to come, and his brother Morrison.

Vigy established its new post at the corner of Fourth and Wood streets. But no sooner was water pumped up than a giant ball of fire closed in on the volunteers, surrounding their location. It was a life-threatening situation, to which the red shirts responded in short order. With no time to uncouple their hoses, they cut them off and dragged their engine and themselves toward Smithfield Street. Vigilant then set up a work station along Smithfield Street and finally at Diamond Alley to protect a row of old wood frame dwellings there.

Sweeping up Second Street, the wind-driven fire fanned out north on Third Street and south across Front or First Street onto Water Street and into the warehouse district. Fires were quickly touched off on building roofs, most of them being of wood-shingle construction. These burning shingles in turn were carried airborne by the wind and landed on other rooftops, setting them afire. Many of these fires could not be seen from ground level due to thick layers of low level-smoke. Many structures were burning long before anyone noticed them.

Along Water Street merchants were evacuating their stores including those whose business was clothing, foodstuffs, paper, household goods, and the like. It was a mad scene in which every available wagon, cart, dray, horse, and human was pressed into service. Running, many times into one another, they made trip after trip from building freight docks to the Monongahela River shore, unloaded, and then went back to get another load.

Over on Third Street, the fire marched up the crowded avenue, torching everything in its path. As it rolled up the narrow canyon formed by opposing building fronts and the street, Mayor William Howard's home was swallowed up by the passing waves of flame. The sight and sounds were frightening. A combination of searing heat, wind swept fire, and crackling flames punctuated every so often by loud explosions. It was an unstoppable wall of fire. People fled their homes and businesses, abandoning furniture and belongings in the street as the flames bore down on them. Even curious onlookers, taking in the spectacle as some sort of entertainment, were routed in fear of the size, speed, and intensity of the rapidly moving fire. A long line of people on foot and families with wagons of all sizes and descriptions crossed the Smithfield Street Bridge to escape the fast moving fire.

Widening its course, the fire spread over to Fourth Street and then Diamond Street by the time it crossed over at Wood Street. Windblown burning coals showered adjacent buildings like so many red snowflakes. Five blocks north to south were now involved, plus the area from Ferry Street east to Wood Street. There were no signs of letting up, in fact, conditions were about to get worse.

Down at the Monongahela River wharf, the shore was packed tight from Water Street to water's edge with merchandise emptied out of the many large warehouses. There were products of every imaginable type. Unnoticed, in all the confusion, flying sparks landed among the closely packed crates, barrels, bags, and boxes, setting them on fire. When finally they were finally discovered, panic set in as people tried to separate the burning items from the non-burning. This only made matters worse, and in a short time the winds drove embers into open spaces, allowing additional fires to start in the stacked and stored wares. Fires sprang up, fed by flammables and other accelerants like coal oil and wax. It was plain to see that the entire levee was beyond help. It was just one great burning mass of flames ravaging anything, everything there. When the fire reached the water's edge, riverboats that were tied up at the shoreline cast off into mid-river, dropping anchor in the safety of the opposite shore near the outlet of Saw Mill Run.

Now, working its way east along Water, Front, Second, Third, and Fourth Streets, traveling from one side of the street to another, the zig-zagging fire gobbled everything in its hungry path. Volunteer fire companies stood by, helpless. Their hoses had long since exhausted what little water remained in the city reservoir. Some companies' hoses were severed by crossing freight wagons. To create a fire break, gunpowder charges were used to blow up buildings in the fire's path. By mid-afternoon, the wind was deflecting the fires north perimeter toward the rocky bluff formed by the corner of Grant's Hill and Boyd's Hill. As the sun rose in the clear spring sky, so did the temperature. The combination of heat from the fire, wind, and the bright sun were like the smithy's forge, one observer commented. In fact it was so hot that zinc roof coatings melted and ran down the rain water conductors.

When the inferno neared the intersection of Water and Smithfield streets, it enveloped the posh five-story Monongahela House Hotel in spectacular fashion, as if it were a hollow shell of papier mache. Spectators gasped as first one wall collapsed out onto Water Street, exposing the burning cross-section of the interior, followed by the entire structure sagging inward and cascading down into a flaming mass of red hot coals.

This collapse created a moving wall of in-

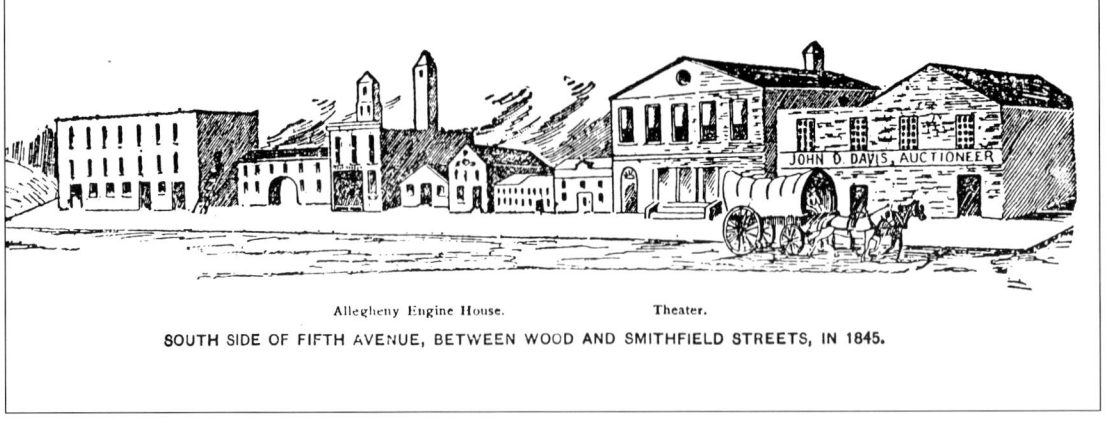

This 1845 view of the south elevation of Fifth Street, between Wood and Smithfield, shows the location of the Allegheny Company's enginehouse, marked by the tall hose-drying tower. (Carnegie Library of Pittsburgh)

In this drawing, Pittsburgh's volunteer firemen can be seen bringing up their equipment to attack the great 1845 fire. Their efforts were largely in vain. (Carnegie Library of Pittsburgh)

The view from Boyd's Hill after the 1845 fire was one of destruction and desolation. This painting shows one of the few recognizable structures still standing, the Third Presbyterian Church, center, at the corner of Third and Ferry streets. (Carnegie Library of Pittsburgh)

tense heat carrying flying sparks and burning fragments that fell upon the covered wooden bridge over the Monongahela River to South Side. Instantly, flames appeared on the old, dry span, whipped up by the unchecked winds in the river channel. Progressively, the fire crawled along the bridge timbers, crossing the river in rapid fashion. Some years later, Judge Thomas Mellon described the sight as "Being like a straw rope on fire." At the South Side end of the bridge, the fire burnt itself out as it reached land. All that remained were the half-dozen blackened stone piers standing like sentinels out in the river.

By late afternoon the line of fire had overrun the blocks between Grant, Cherry, and Ross streets, consuming the Scotch Hill area, crossing the canal, and pinching in along the narrow community at the base of Boyd's Hill. Locally known as Pipetown, a plan of lots laid out by clay pipe manufacturer Billy Price, residents scattered as the fire flashed through the homes and factories of the district.

Among the major industries wiped out by the fire were the Bakewell, Pears & Company glass factory and the Dowlas Iron Foundry. By seven o'clock, the wind had subsided and the fire, with no force or fuel to feed it, burned out on the river shore near what is today the Tenth Street Bridge. Pittsburgh's greatest fire was over, but the impact of the event was just about to be felt.

At first light on Friday, April 11th, the fire ravaged city awoke to a silent world cloaked in gray ash. A thick layer of damp, smoky haze blanketed

> **In the harsh reality of daylight, the magnitude of the disaster was revealed.**

the entire valley of the three rivers. Nothing was stirring. It was a ghostly quiet scene, devoid of animation, like a still landscape painting.

For Pittsburgh and its people, it had been a restless night. Exhausted from their frightening ordeal, many citizens slept out under the spring sky and stood guard over their chattels to prevent looters from making off with them. There was no light to see by, save the occasional lantern or small fire built to keep the night chill away. Many buildings were in danger of collapsing due to structural fire damage. In this darkness were people taxed beyond their physical and mental capacities. The coming dawn would only add to their distress.

Slowly, the smog began to lift, revealing a devastated city. Looking from Grant's Hill, to the old forts was a study in contrasts. On the right, or north end of the triangle, the city remained untouched, seemingly oblivious to the tragedy visited on the remainder of the community. On the Monongahela River side it was a very different scene. A 24 square-block area from Pipetown to Ferry Street and reaching as far north as Diamond Alley was leveled, reduced to a charred mass of blackened timber and twisted debris, brought to earth by the might force of the hellfire.

Only a thick carpet of ashes, broken by the occasional stone foundation wall or wobbly chimney remained of the 900-plus buildings that fell victim to the blaze. There was a penetrating odor of afterfire that burned the nostrils and made breathing difficult. Life was scarce, only an occasional person or two standing in the streets, stunned, looking at the remains of what 24 hours earlier had been their home. Others walked around in a daze. Here and there a stray dog would appear, disoriented and lost. In the harsh reality of daylight, the magnitude of the disaster was revealed. To the individual the devastation seemed complete; to the city it was the very closest of brushes with the grim reaper. Remarkably, only two people died.

Statistics told a story of major damage and widespread destruction. Nine hundred and eighty-two individual buildings had been obliterated, reduced to dust. Among the more prominent addresses lost were, in addition to the Monongahela House, the *Daily Chronicle* newspaper offices, the Merchant's Hotel on Wood Street, both the Associated Reformed Church located on Fourth Street and the Baptist Church, plus Western University, Custom House, and Philo Hall where the mayor's offices were quartered. Ironically, the Fire Navigation, Fireman's and Penn Insurance company offices were also laid waste by the fire. Only three buildings were left standing in the city's second ward.

Even the Vigilant Volunteer Fire Company was not spared, its engine house consumed by the wind swept demon. Lost too, when their building went down, were several lengths of pirated fire hose. After a fire, it was common practice among the volunteer companies to appropriate the other fellows' hose sections. As a result, companies began to stencil their names on the side of the hoses but it did little good. Any hose was fair game, and the lettering was later filed off during "filing parties," that were fueled by large quantities of spirits. But for now, Vigy's hose reel would not be any longer.

Estimates set the total loss in dollars at $2.4 million minimum, of which $900,000 was personal property and $1,500,000 was real estate. Some would claim $8 million was closer to the actual figure. In fact, one third of the city's geographic area, approximately 56 acres, was wiped out. Two-thirds of Pittsburgh's industrial and commercial wealth was taken by the fire. Losses were more personal to those 12,000 residents who lost their homes and practically everything else. Many had escaped with just the clothes on their back. For them there was nothing to go back to, no homes, no neighborhoods, not even streets in some cases. All that was left was the endless sea of grey and black covered earth.

Property insurance holders were few if any. Published individual losses listed James Crossan's $35,000 claim for his Monongahela House down to the $15 sought by servant Anne Collins. A total of $79,000 in claims was paid out of the original total of $870,000. Unlucky policyholders were those who were insured by Pittsburgh companies. All of their financial assets went up in the smoke of the fire.

With it all there was a miracle of sorts to come out of the awful tragedy. Despite the enormity of the fire and the heavy property loss, only two human lives were known lost in the conflagration. Early in the ordeal, Attorney Samuel Kingston left his residence near the corner of Third and Ross streets as the fire approached. For whatever reason, he returned home to rescue his hand-painted piano and, in the confusion, he entered the basement of a dwelling two doors away. His incinerated body was taken from that cellar early in May. The other known fatality happened a few blocks away. On April 22nd the charred bones of a Mrs. Maglone were discovered in the foundation of a building at the corner of Grant and Second streets. Some residents remembered her walking in that area wearing a colorful flannel dress and bonnet. Why she was in the building was not known.

There were also two scientific discoveries that came about as a direct result of the fire. Freeport chemist and physician Doctor David Alter came across a shard of flint glass in the remains of the Bakewell Glass Factory. Shortly afterward, he ground a prism from the piece. This eventually led to his 1853 discovery of spectrum analysis. From McKeesport, Squire Wampler traveled to Pittsburgh to collect his souvenir of the landmark fire. Driving by wagon to the Monongahela River shore, he crossed over by ferry and started his search at the Bakewell site. Wampler took his piece of flint glass home and, along with another piece of glass, he had a telescope built. It was thought to be the first telescope west of the Allegheny mountain range. Wampler toured up and down the Yough and Mon Rivers, stopping at each town to sell looks at the stars. One of his patrons was a nine-year-old Brownsville boy named Alfred Brashear, whose fame as a star gazer was still many years off.

Of first concern was the plight of families made homeless by the fire. In quick fashion, Mayor William Howard and the City Government made provisions to house them in public buildings, including the courthouse. A public appeal was made to those who were lucky enough to have escaped the fire to help their less fortunate brethren. Most of the needy citizens were impoverished or held, at best, menial jobs. Their homes, situated in the original older section of the community, were the first destroyed by the flames. It would not be possible for them to survive, let alone get back on their feet, without help.

As word spread beyond western Pennsylvania about the catastrophe at Pittsburgh, an immediate show of support was forthcoming from all points of the compass. The neighboring states of Ohio and New York sent monies totaling $10,081 and $23,265 respectively. Massachusetts was the next largest out-of-state contributor at $16,741; Indiana the smallest at $52.

Other states that answered the plea for help were Maryland, $11,513; Kentucky, $5,773; Louisiana, $7,167; District of Columbia, $2,872; Delaware, $1,322; Tennessee, $1,259; Alabama, $1,652; New Jersey, $557.96; Michigan, $100; Mississippi, $1,291; Georgia, $470; and New Hampshire, $329. There was even a $651 gift sent from Rothschild brothers in Europe. At the request of Cornelius Darragh and William McCandless, the Commonwealth of Pennsylvania earmarked $50,000 to cover emergency costs and enacted special legislation to ease or forgive state and county taxes through the year 1849. It should be noted that the City of Pittsburgh contributed no money to the cause.

In addition to the cash, several nearby communities donated food to victims of the fire. A barrel of sauerkraut was sent from the community at Economy. From Wheeling, West Virginia came 3,000 pounds of fresh bacon and 100 pounds of flour. Fifty-eight bushels of potatoes were sent by the town of Meadville, Pennsylvania.

Almost immediately, the city began to rebuild itself. A major effort was initiated to clear away the damage so that construction could begin anew. City newspapers carried the story of Pittsburgh's reconstruction far and wide. They told in detail of

William Howard was the mayor of Pittsburgh during 1845. To him fell the awesome responsibility of coordinating local efforts to rebuild the city and get Pittsburghers back on their feet. (Carnegie Library of Pittsburgh)

the fire's destruction but sounded positive notes for industries still in business and the prospects for those rebuilding. Pittsburgh citizens were portrayed as patient, hardworking people, dedicated to the rebirth of their city. They served as a shining example to all and were one of the major reasons that lending capital was made available to the struggling town of 30,000 souls.

Good weather prevailed during the early days of the reconstruction process. In short order demolition work commenced and the debris was cleared away. New construction work sprang up, much of it on existing foundation walls. A few establishments, like the Eagle Livery Stable and Johnston & Stockton's book shop, were under roof within a week. By mid-May, several dozen homes were framed in and several hundred foundations were underway. The entire town was a beehive of activity. Citizens and workmen alike moved stacks of lumber and drove wagonloads of stone and brick, all putting forth maximum effort for the singular purpose of re-establishing their city.

It was the beginning of Pittsburgh's greatest expansion period to date. Local population grew, due in some measure to the influx of tradesmen who came to the city to find work and then stayed. Some residents decided to relocate farther east, helping to populate areas beyond Grant's Hill and the Northern Liberties. Real estate values were on the rise. Technology, too, took a big stride forward when John Roebling's wire rope suspension bridge design, the first in the world, was built on the stone piers of the old Mon bridge, itself a fire casualty.

For some individuals, the fire's aftermath was a springboard to wealth and success. One of these opportunists was fledgling attorney Thomas Mellon.

For some individuals, the fire's aftermath was a springboard to wealth and success.

After the fire, he built 18 rental homes which returned double digit profits on his original investment. Those funds became the first building blocks that led to the Mellon family fortune. Quoted later, when he was Judge Mellon, the financier remembered that "Instead of depression, it gave an impetus to every kind of business. There was ready employment at better wages, new life and increased value was infused into the real estate."

While the city was occupied with recovery, so were its volunteer fire companies. In dire straits was old Vigilant which had no home, it being lost in the ordeal. But the resourceful Vigys found a solution to their problem, the end result of which would prove to be humorous.

After the great fire, the brigade pulled its engine into open space at the Diamond Market House. When the butchers set up on sale day, they pushed the old machine back outside. Vigilant's management didn't like their engine being exposed to the elements and told the merchants so. Their words fell on deaf ears. The pumper continued to make regular trips outside no matter what the weather.

Many market patrons were curious about the contraption and frequently asked what it was. Annoyed by the many inquiries, the merchants would reply "It's a sausage stuffer." This was always good for a chuckle from the passers-by, much to the annoyance of the red shirts.

To solve the dual problems of outside storage and being the butt of jokes, a company sewed muslin cover was fitted to cover the entire machine. Only the lower wheel quadrants and tow rope were visible. When the fire call sounded, off went the engine at a dash, looking for all the world like an elephant. Unsuspecting citizens discovered too late that the joke had been on them. As the engine raced by, the cover was removed to reveal Vigilant's fire engine.

On May 3, 1845, Eagle Company offered part of its quarters to house Vigilant. The offer was graciously accepted, with thanks, and a company of Vigy members made a visit to Eagle's headquarters to acknowledge it. Once again, lodge brothers were helping each other in a time of need. The Market House butchers again had the place all to themselves, and Pittsburgh's "elephant" roamed the city streets no more.

With the ordeal behind them, the volunteer companies now had time to think and evaluate their performance at the monumental fire. After much thought and discussion, the consensus was reached that important lessons had been learned. It was education of the most brutal type. Better firefighting equipment and more of it was needed. More and better-trained volunteers were a must. Water supply lines and distribution facilities would have to be improved and increased. There also had to be a better way to compensate firemen than the pay-per-fire system then in existence. Many times this had led to competing companies fighting and arguing among themselves while someone's property burned away.

One of the positive developments to emerge from the investigation was the formation, in August 1845, of an organization called the Firemen's Association of Pittsburgh and Allegheny. Its stated purpose was "promoting good order, efficiency and harmony in the fire departments of the twin cities." This fraternity was responsible in large measure for improving the capabilities and quality of the firefighting services.

In the late summer of 1845, Neptune's brigade was busy trying to rebuild itself into a viable firefighting company. Its engine had suffered significant damage while in service at the Second Presbyterian Church fire. In addition to equipment problems, the blues could muster only feet of usable hose out of the 915 feet they originally had.

Unfortunately, there was no company treasury money to finance the work. Help did come, however, from an unexpected quarter. A benefit performance for Neptune was given by the S. P. Stickney Circus to raise money for the repairs. This generosity enabled Neptune to replace the lost hose plus sign a December 10, 1845, contract with the Sheriff & Co. Engine Works to reconstruct its pumper.

Early in the new year of 1846, Allegheny Company moved into new quarters on Fifth Street. In addition to more space for their machine, the building provided easier egress out onto the wider thoroughfare. Shortly after the move was completed and the volunteers were settled in, a new slate of operating officers was elected.

Neptune's blue shirts had cause for celebration when, in July of this year, they took possession of their refurbished engine returned from the contractor. "Better than new," some said, remarking about the fresh coat of royal blue paint and the highly polished brightwork. The bill tendered for the work was $690.87. To properly display the restored engine, Neptune's management, acting on a motion by Member John D. Bailey, exhibited the machine at the next civic parade. Dressed only in the stars and stripes, big Neptune, a model of patriotic simplicity, passed by to receive the warm applause of those looking on.

Pittsburgh's volunteer fire companies continued to exhibit their social and fraternal presence at public events, reminding all citizens that they did more than fight fires. No matter which of the volunteer companies was involved, there was always that competitive spirit between the brigades, a fierce show of both personal and company pride.

Racing of all types was highly popular during these years and the firemen had their own particular brand of it. Usually, two competing companies would pull their engines along a predetermined race course, the first to cross the finish line was the winner. Simple enough you say, but there was always an element of surprise or personal interpretation of the rules to make things interesting. Take the famous or infamous fox tail race, for example.

The competitors were Vigilant and Allegheny. The race course was up Wood Street to Liberty, down Liberty turning onto Wayne, then out Penn Street. A right turn at Hand Street would bring the contestants back to Liberty, From there a final run down Liberty, turning at Market Street, would end up with the last lap on Fourth. The prize: a red foxtail belonging to and tied on to the hosereel of Vigilant.

At the drop of the starter's flag the race was on. Things progressed nicely with Vigilant leading out with a one-block head start. Within a block, Allegheny's tan shirts struck out after the prize fur. All went according to plan with Vigy maintaining the lead through the corner of Hand Street and Liberty. As they came down to Wood Street, person or persons unknown directed Vigy down that route; Allegheny straight on, changing engine rope pullers at Sixth Street. Vigilant, now unable to call upon their relief men at Liberty and Market Streets, had to rely on their original team for the entire distance.

As it turned out, the detour proved to be of little consequence, save some shortness of breath, as Vigilant won the race handily. *Red Bird* pulled up at its enginehouse, with hose reel behind, foxtail flapping in the breeze. A disappointed Allegheny came along side of the red shirts, a picture of dejection. One frustrated member of their group plucked the tail from Vigy's reel and hightailed it out of there. Vigilant lost the foxtail but won the big race. Locals talked about nothing else for months after. It was remembered as the greatest fire engine race ever between two Pittsburgh volunteer fire companies, rivaling in folklore the legendary duels between the riverboats *Natchez* and *Robert E. Lee*.

On into 1847, the fire brigades continued to place themselves in the public eye. Whatever the occasion or chance for notoriety, the boys seldom missed an opportunity to strut their stuff. Particular favorites were parades, of which seemed to be prompted by any reason or event. Firemen were as loyal to the parade call as they were to the fire call.

At noon on September 1, 1847, a large procession marshalled to honor the local visit of George M. Dallas, vice president of the United States. Among the groups assembled were several volunteer fire companies, complete with their decorated engines, hose reels, and uniformed membership.

One by one, the parade units passed by the flag-draped Wood Street reviewing stand. Eagle was the first to go by and did so stonefaced, with eyes fixed front, giving no acknowledgment whatsoever to Dallas. Unperturbed, the vice president turned to greet the men of Allegheny as they approached. On signal, their rank and file, together, tipped their tan hats in salute, a gesture that Dallas returned with a smile and a nod. In turn, Vigilant's company came into view. Their entourage included, in sequence, a hose reel pulled by four horses, the firemen's brass band, the engine nicknamed *Red Bird* drawn by a matched team of six horses, plus 80 marching members attired in rich red uniforms. Fourteen floral wreaths, presented by the ladies of the city, decorated the equipment.

When the group came parallel with the dignitaries' platform, Vigilant's Captain David Campbell gave the command "Front Face, nine cheers for George M. Dallas, Vice President of the United States." A mighty roar of voices rang out with the tribute, echoing off the surrounding buildings. It was perfectly clear that Vigilant would have no part of the political prejudice, a fact noted by both Whigs and Democratic members in attendance.

Neptune's early engine had seen such hard use by November 1845 that it required about $300 to repair it. This document requests the governing councils to appropriate the money. (Library and Archives Division, Historical Society of Western Pennsylvania, Pittsburgh, PA)

> Although united in purpose, Pittsburgh's volunteer fire companies were as individually different as the colors they wore.

Shortly after that event, the Vigilant received notice that the property it was renting as enginehouse and headquarters was sold. Noted businessman Charles H. Paulson, a local hat merchant, purchased the old building with an eye to razing it and constructing a new and larger shop. Although the sales agreement stipulated that the work would not begin until the firemen found new quarters and vacated the premises, both parties received a surprise on the following Saturday.

Vigilant members were returning from yet another parade, and when they came in sight of the enginehouse lo and behold there were workmen on the roof dismantling it. The big red engine hadn't come to a complete stop but already the volunteers were on the ground, scurrying up the contractor's ladders and confronting the laborers. There was a brief scuffle, highlighted by a flurry of flying bricks and lumber, sending the workmen back down their ladders on the run. Immediate plans were made to relocate the brigade. Eagle once again offered temporary quarters. In a conciliatory gesture, Charles Paulson offered the use of one of his other properties until a permanent home could be built. The company finally wound up in a Water Street address, owned by John Guthrie, known as the Custom House.

On March 20, 1848, Vigilant's officers met at Eagle Hall to discuss plans for a new enginehouse. Joseph C. Kear drafted the construction drawings and wrote the specifications for labor and materials. These in turn were handed to the company building committee consisting of David Campbell, B. C. Sawyer, Jr. and James Irvin. Vigy would finally have a home to call her own.

In that same month, while meeting at Eagle Hall, the reds planned their participation in an upcoming parade to honor Henry Clay on his first visit to Pittsburgh. Charles A. Crosby, James Petrie, John Liggett, N. P. Sawyer and William Alexander served as a Committee to plan for the event. It was a bipartisan effort supported by both the Democratic and Whig members. A grand torchlight parade, complete with fireworks, greeted the well known statesman and orator. Earlier that day, he stepped down the gangway of the boat from Brownsville. He now took the reviewing stand as night extinguished the last rays of daylight.

There turned out to be one additional reason for Vigy to celebrate this day. While the members paraded with their equipment, the Custom House, where they were quartered, caught fire and was totally destroyed. If it had not been for the Clay celebration, the red machine would have been inside and most certainly, lost. It was truly a day for all to be thankful.

On the 7th of May, Vigilant assembled its members for duty for a somber and personal duty. One of their own, Fireman H. Bellas Allward made his final journey home to Pittsburgh, a casualty from the war in Mexico. Allward had no living relatives, no family but Vigy. It was the members sad but proud duty to carry him to his place of last rest. David Campbell, R. Biddle Roberts, and John Liggett purchased the grave plot and made the arrangements.

Despite their sadness, fellow members recalled, a funny incident, typical of Allward's devotion to fire duty, in fact, a model for all firemen regardless of affiliation. While fighting a fire in a building on Water Street one night, bystanders called out to Vigilant's crew that a child was trapped inside and crying out for help. Without hesitation, Allward dashed into the fiery structure, reappearing a short time later with a very large tom cat cradled in his arms! A rousing cheer went up from the crowd assembled. Human or animal, Bellas Allward was a man devoted to his duty as fireman.

Although united in purpose, Pittsburgh's volunteer fire companies were as individually different as the colors they wore. Yet, there were still many similarities that existed, held onto since the lodges were first chartered. One of these common threads was a love of fistfighting. Although the practice seemed to diminish from 1830 to 1840, it intensified during the period 1840 to 1850. This was the decade of the so-called "Engine Wars," a time of fierce competition and intense rivalries among fire companies.

Some notable exchanges took place between the two elder statesmen of the group, Eagle and Allegheny. Eagle liked to flaunt its title as the first organized fire company. Allegheny always had difficulty coping with the reality of being second, especially when reminded of the fact by Eagle. To compensate, Allegheny's engine was built twice that of Eagle. Taunts and barbs were frequently tossed by men of both groups in an effort to instigate a fight.

One of the many joint provocations happened on the way back to the engine house after a fire call. Returning on the run, men and engines pulled abreast and even with each other and began to jibe back and forth. Verbal insults were liberally traded and there was much hand waving and posturing on both sides. One of the Allegheny chaps, a big strapping lad who was as wide as he was tall, tripped and fell to the ground. Both lodges pulled up by the fallen fireman. Suddenly, one of Eagle's red-shirted clan, with a mouth larger than his five-foot frame, squeezed into where the downed man was lying and in a loud and boisterous voice, said "There, you might have known I'd have knocked you down if you looked at me that way." This outburst brought the injured man to his feet, exclaiming "You! You knock me down? Why you little runt, I could eat you for a quail on toast and swallow your engine for dessert." The exchange brought an instant roar of laughter from both groups. Humor saved the day and sent each company on its way home without further incident.

Neptune's company, in addition to fire calls, was readying a celebration of its own. Some 19 of the men in blue answered the call for duty during the war with Mexico and seven of those souls perished in the conflict. In July 1848, the remaining 12 returned to Pittsburgh and their beloved Neptune. To properly welcome home these firemen veterans, the steamboat *Pennsylvania* was chartered by the officers and membership. Cruising down the Ohio River, the delegation met the returning heroes at Beaver. Someone decided that a salute was in order and fired a cannon from the deck of the ship. Unfortunately, sod was the only wadding available and its resilience caused a recoil that pitched the gun overboard and into the river. Company members pledged funds to make good the loss. More importantly, all Neptune brothers were reunited with their lodge fellows and their families.

No sooner did Neptune members get settled in after their celebration than a notice came ordering them to vacate their enginehouse and meeting hall. Trouble had been brewing for some time with the nearby Smithfield Methodist Church. It seems the boys became rowdy on occasion and ignored the requests of church elders to settle down and quit this type of behavior. The 'True Blues' were a lively group with many young, athletic members. Every sporting event was followed a spirited social event at which liquid refreshments were served into the wee small hours. A particular favorite was the ritual of toasting their boating club which rowed the Allegheny River in their eight-oared barge *Fashion*.

During August, Vigilant also went down river by steamer to welcome a group of Cincinnati, Ohio, firemen visiting the twin cities as guests of Uncle Sam Company. Low water grounded the craft about 10 miles down the Ohio forcing men and machines to be off loaded to complete the journey to Allegheny City by land. A parade through the streets of both cities welcomed the visiting firemen, the locals providing a warm greeting to the Ohioans. When their engine passed during the parade it caused a stir. It was a type never seen before, a box or barge-like affair in which the firemen sat and, with a rowing motion, worked the gang levers that pumped water to the hoses. Onlookers were quick to nickname the craft "Rowboat."

Arrangements for the parade through Pittsburgh were handled by Vigilant, a fact appreciated by Uncle Sam. To show gratitude, Uncle Sam presented silver name plates engraved with the word "Vigilant" to their benefactors, thanking them for what they had done in the name of brotherhood. These were presented to Vigilant at a meeting in the headquarters of the Uncle Sam Fire Company, Allegheny City, on September 6, 1848. Sixty Vigilant members were in attendance.

This same month, the City of Pittsburgh sold its Cecil Alley water pumping station for $24,000. Allegheny City was also constructing its first municipal water system at this time. There was a definite need to improve water supply service to the growing community. Up until this time the only means of supplying water was by well or the river. Delivery was made to individual homes and businesses by a corps of five dozen hand-propelled water carts.

Objections to the central water supply system were raised by those who made their livelihood supplying water. But progress was not to be denied. Work continued on the project to bring it to a successful completion. Most Allegheny citizens knew that in order to prosper like their bigger twin across the river, a modern, functioning water supply system was a must. Improved water delivery would also help the local fire companies, which were trying to protect an ever-growing landscape.

With the arrival of autumn, parades and pageants were a weekly occurrence in the city. Almost any event or holiday was cause for a celebration. Of course, all the local volunteer fire organizations were in the forefront of organizing and, above all, participating in these public processions. Parades were a universal favorite of all firemen, ranking right up there with eating, drinking, and sporting events. The fire brigades were

The 1849 petition to the Committee on Fire Engines and Hoses from the Fireman's Association to purchase a new fire engine for the Good Intent Fire Company. (Library and Archives Division, Historical Society of Western Pennsylvania, Pittsburgh, PA)

Built by the Duquesne Engine Works owned by James Rees, this steam-boilered fire engine was owned by the Eagle Company. It was the first of its type west of the Allegheny Mountain range. It was also the first to be drawn by a horse team. (Carnegie Library of Pittsburgh)

important elements of any organized marching display, ranking second only to the local fraternal and labor organizations. Many times the fire engines were horse drawn, brightly decorated with floral arrangements and numerous flags. Starting on Liberty Street, between Hay and St. Clair, the companies stepped off after being inspected by city fathers. Parading in with the regular uniformed marchers, other members would dress up like revolutionary soldiers or Indians.

Vigilant meanwhile was still housed in temporary quarters at Patterson's Stable. When the new enginehouse space was finally renovated, there was cause for celebrating and, in fact, an afternoon parade was assembled with all the volunteer units participating. By evening, attention turned to the program playing at the Old Theater. Most firemen and their ladies were in attendance. Before the last performance was over, the mood turned from joyous to somber. One of Neptune's company left the theater at intermission and accidentally severed an artery. Despite immediate, medical care, the bleeding could not be halted and he died within a short time. Two days later, the fire brotherhood from both cities gathered again, this time to conduct the funeral procession for their fallen brother.

Later in January, Vigilant members finally moved out of the old livery premises, taking everything, especially their beloved *Old Red Bird* engine. They proceeded to set up shop in the new building, organizing and storing the contents of the many boxes and bundles brought from the old building. Monday, February 5th dawned cold and rainy, a typical Pittsburgh winter day. A fire broke out at the Evans Flour Mill, the work of an unknown arsonist. The wooden structure was located at the corner of Water Street and Redout Alley. In no time, the entire frame was engulfed in flames. A stiff breeze blowing in off the river fanned the sparks and firebrands eastward, across the alley, setting several of John Irwin's row houses ablaze.

Remaining unchecked, the fire continued to burn eastward along Water Street. Vigilant's brigade, not yet operational, hurried to move their engine out and head for the fire scene. Someone in the rank and file remembered that Vigy's hose was in storage at James Irvin's store and headed there to retrieve it. Meeting their engine at First and Ferry Streets, Vigilant's hose reel immediately began to play out their lines. Allegheny City's Uncle Sam unit arrived and promptly connected a hose to a nearby water plug and then ran a line to Vigilant's engine, supplying it water. Minutes after, Eagle Company pulled up and began to engage the flames. Firemen donned their rain capes as the downpour intensified. The major effort was concentrated along First Street, behind the burning structures, in an attempt to cut off the fire and prevent it from spreading into the city.

Commanding the group was veteran Captain David Campbell. It took several hours of grueling work to extinguish the fire. The steady rain, driving wind and falling temperatures hampered all efforts and took their toll on the men. Campbell contracted a respiratory illness that confined him to his home for two months. Other company members were exhausted from the ordeal, barely getting their equipment back in the enginehouse. Still, they were all better off than Eagle's Captain William Hays. He was the lone casualty caused in the fire, losing his life when he entered the burning residence of John D. Davis to see if there were any residents trapped inside. Finding none, he attempted to exit the burning structure but was crushed to death when the walls and floors fell in.

Summer also brought its share of suburban fires. For the second time in four years the Western University of Pennsylvania campus burned on July 6, 1849. The first loss was due to the great 1845 epic. After that, the school trustees decided to rebuild the school on Duquesne Way and, when it was completed, took out $8,500 worth of insurance to cover the new structures and equipment. Nobody could have known that this policy would pay off in such a few short years. The entire school, its buildings and equipment were a complete loss. At the time, there was some public grumbling that additional and better fire equipment might have cut the losses. This point would be resurrected at a later date to support requests for fire equipment purchases.

One of the darker aspects of local firefighting came to pass 10 days later. Historically, since the earliest days of the organized fire fighting companies, each brigade was paid a fee by the municipality for each fire call answered. At times, payments fell behind but, eventually, all claims were paid in full.

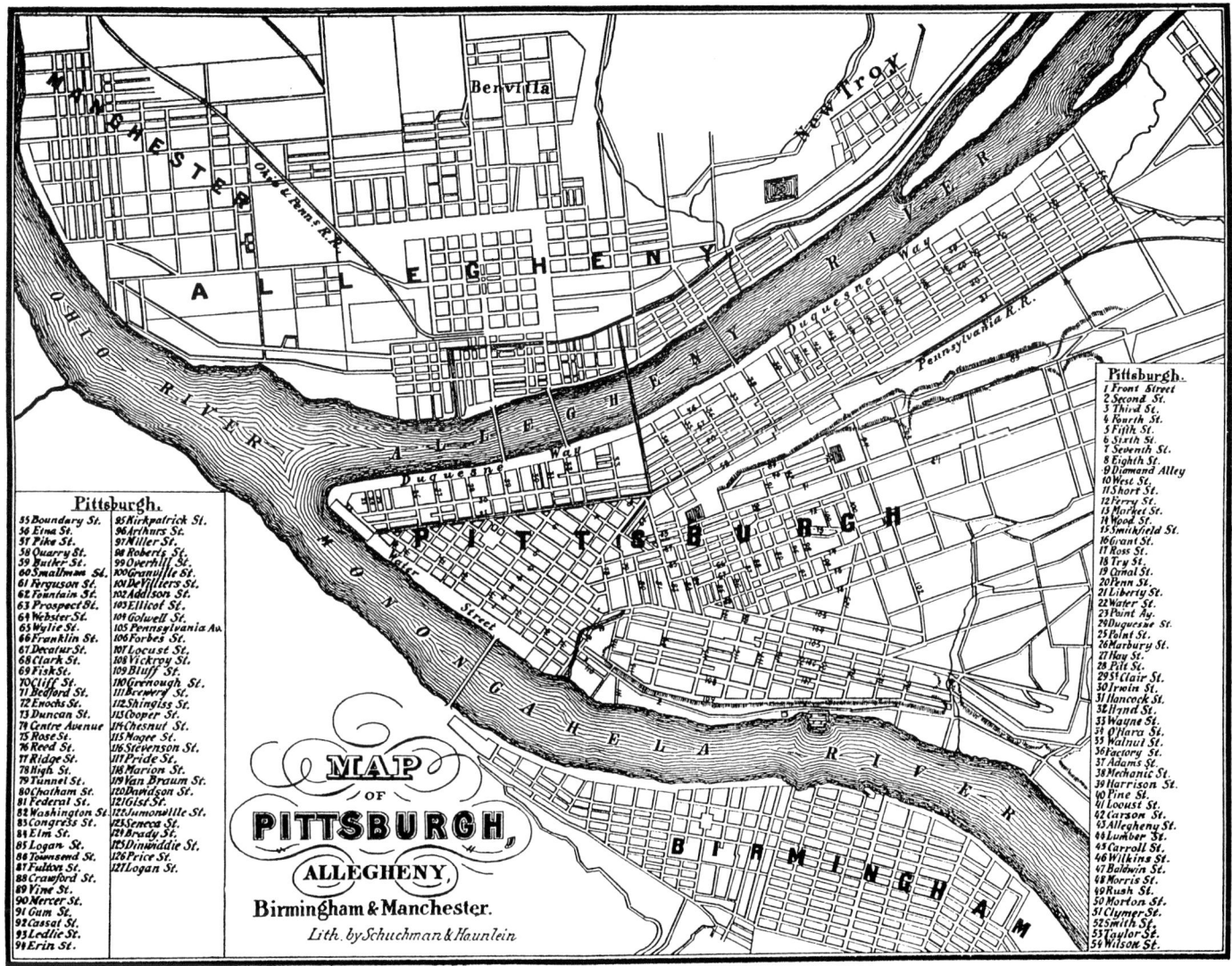

There was marked expansion of Pittsburgh and its cross-river neighbors in the first half of the 19th century, as shown in this 1851 street map. (Carnegie Library of Pittsburgh)

It was nevertheless annoying to the firemen, usually a necessity to the city treasury. Still, the practice continued and in some instances became abusive.

Firemen of Allegheny City had long been pleading with their local councilmen for back monies due; for whatever reason, their requests fell on deaf ears. In a final effort to get paid, the fire companies of Allegheny banded together and submitted a written ultimatum: "Either pay up or a work stoppage will be called." Still there was no response.

Now turn the calendar page forward to July 16, 1849. The fire bell sounded for a fire in the carpenter shop of Henry Charles on Gay Alley, between Arch Street and the Diamond. Responding firemen found their striking counterparts armed and blocking access to the fire scene and water sources. Willing volunteers, including Manchester Borough's Lafayette Company, stood by helplessly as the flames ravaged first one building and then another. Seeing the column of smoke from across the river, Pittsburgh fire units mobilized and started for the cross river bridges. To their surprise, they, too, were turned back by club swinging firemen. It was an ugly scene. Against a backdrop of burning buildings, firemen brawled with each other using fists, throwing bricks and swinging bats. Any attempt to lay hose met with immediate and brutal consequences, not to mention cut-up hoses and engine damage.

The event was a disgrace played out in the national newspapers. One report called Allegheny City "The only city in the United States where firemen stood by cheering the progress of the flames they were sworn to extinguish." In all, 35 buildings, the Presbyterian Church among them, were reduced to ashes. Estimates set the total loss at $75,000. Those who lost property not only had that burden to suffer but also were penalized with increased property insurance rates. Eventually, the strike instigators were brought to court, convicted, and fined. But it was too little, too late for those whose property was destroyed, and the same held true for the good name of the majority of Allegheny City firemen and the image of their city.

Back across the river, Pittsburgh's volunteer firefighters had their own problems, one of which was an increase in the number of false alarms turned in. The ratio of bogus reports to actual fires was about 2.5 to 1. In the year 1849 alone, volunteers fought 46 legitimate fires. During that same time period the number of counterfeit alarms reached 116.

Arson was another serious situation, also increasing in frequency. Over time, some fires fell into a pattern, like the same type of building or the same neighborhood. Fire calls at night, over weekends, or on holidays were immediately questioned. Responding firemen could almost guess which fires were torch jobs just by the address reported. Neptune Company had serious concerns after responding to five false alarms in one night. Suspicious, they played detective in several alarm-prone areas, catching three firebugs lighting off the straw in Matthews Livery Stable along Penn Street. Their handiwork resulted in a 30 year jail sentence, an example of what punishment awaited others who liked to set fires.

Passing the mid-point of the century, Pittsburgh saw its borders continue to expand. A combination of much-needed room for new citizens to settle, plus the exodus after the 1845 fire, led to the establishment of a number of new communities and boroughs. Two miles up the Allegheny River from the heart of town, the borough of Lawrence, later Lawrenceville, was a growing center of new home construction. Across the river, Allegheny City was flourishing. The main plateau, known as first bank, had many commercial establishments and fine homes. On the hill above, or second bank, Western Theological Seminary and the penitentiary mingled with residential areas bordering the public commons. Two miles to the west and down river, the new community of Manchester was beginning to sprout. Going east along the Mon and past Pipetown, an area known as Scotch Bottom was building up. Across the Monongahela River and two miles east of Birmingham was another new village called Sligo.

Municipal services, notably fire protection, were unable to keep pace with the growth. There was always a need for additional volunteer firemen, but the real deficit was the lack of modern equipment. Not only in quantity, but in technical capabilities and refinements, the engines and their support apparatus were becoming more obsolete with each passing day, each fire fought. The crux of the matter was the need to pump additional water volume at higher pressure through the hose nozzle. Human power couldn't keep up with the demands of fighting fires in larger and higher buildings.

During 1851 alone there were several complex fires that tested the volunteer's capabilities. A January 25th fire on the Mechanics Bridge posed the problems of working in a narrow space with the possibility of structural collapse. More importantly, attempts to draw water from the frozen river proved to be futile. The Fifth Presbyterian Church fire on March 6th was a series of nightmares.

Shortly after midnight, the alarm was struck. Good Intent's brigade rolled up to the Smithfield Street scene, set up, and tried to pump water, but to no avail. Upon closer examination, the hose nozzle was found to be plugged up with paper and leather wadding! A prankster's joke no doubt. Meanwhile, Neptune opened its enginehouse doors to find their engine's wheels, levers, etc., bound up with rope. Unraveling the mess cost the Blues precious time. By the time they arrived at the fire scene the church was lost. One postscript to the event: While returning to their enginehouse, Duquesne Company's squad was physically attacked by unknown assailants. One member suffered serious injuries.

Local firefighters suffered a more personal loss in March when Eagle's headquarters burned down. Fortunately, members J. K. Perchment and William Wilkinson were in the vicinity and, with great physical effort, were able to pull the engine and hose reel to safety. No sooner had the rear engine wheel cleared the house door than the old frame structure, a mass of fiery lumber, collapsed. Insurance calculated the loss at $800, including 500 feet of leather hose that had been hung up to dry. Damage to the engine necessitated major repairs, which put Eagle out of service. Good neighbor Vigilant offered to share space with Eagle until the company could return to duty.

It was the third significant fire during the month of March, taxing the volunteers to the maximum. The March 4th fire at the old Fountain Inn on Webster Avenue between Crawford and Tannehill started off the month. This was another night fire, recorded at midnight, and very difficult to fight due to close quarters. Needless to say, all volunteer firemen were glad to see March drop off the calendar. But it did not exit without one serious accident happening to Vigilant. Enroute to the Fountain Inn fire, two of the members were run over by their own engine. March 1851 would be a month the firemen would rather forget.

Men and machines were again put to the test once again at the St. Paul Cathedral fire on May 6th. It was a very difficult fire to fight and bring under control. Situated at the corner of Grant and Fifth streets, it was located on an elevated plot of land about 20 feet above street level. This condition had been created by a regrading of Grant Street four years earlier. By the time the volunteers reached the site and deployed, the fire had a head start. And being located on main thoroughfares of the city didn't help matters either. Crowds of onlookers had to be pushed back in order to let the firemen operate their equipment. The flames were well entrenched in the frame structure and several times the fire was thought to be out only to re-erupt once again. Finally, after a five-hour ordeal, all that remained was a charred old hulk that had once been a beautiful house of worship.

During intervals between fire calls, volun-

No sooner had the rear engine wheel cleared the house door than the old frame structure, a mass of fiery lumber, collapsed.

teers spent any time not needed by job or home at the firehouse. There was always hose to be sewn or equipment that needed repairs. Eagle in particular had a major project repairing its damaged engine. Cramped space in Vigilant's headquarters made the task more difficult. Some of the work could be taken outside on nice days to get elbow room and take advantage of the natural light. No matter how close the Reds and Greens came together in the cramped space, they did get along in splendid fashion. This harmony was evident even in the behavior of Vigilant's mascot, the dog "Rolla," but it was not always so.

Rolla was a large bulldog with gray and black markings. He was the property of Vigy member Andrew Fulton, Esq., but was adopted by the red shirts as their mascot around 1850. Rolla loved the Vigilants and they loved him. There was only one thing for which he held a greater affection, and that was the big red engine. In his mind, he was to guard it at all times, whether it was in the enginehouse, on duty at the fire scene, or on the journey to and from. Weather or time of day or night posed no obstacles. When the alarm struck, Rolla was at the ready, poised to leave with his company.

On a return trip from answering a call in Allegheny City, Vigilant was joined by Eagle. At that time Eagle also had a mascot, who had a nasty habit of coming up behind Rolla and giving him a bite on his hind quarters. This went on for several blocks until the companies separated at Liberty and Market streets. Vigilant headed up Market and Eagle down Liberty, except for their dog, which continued to badger Rolla. The snapping and snarling continued till the intersection with Fifth, where the bulldog decided he'd had enough.

With a sudden turn and teeth bared, Rolla attacked his tormentor, jumping on him and knocking him to the ground. The fight intensified, accented by loud barking and growling. Each company urged its respective mascot on. Soon, a crowd gathered to watch the spectacle, and shortly the audience doubled in size. It wasn't very long until Mayor Joseph Barker appeared and asked the respective company captains to call off their dogs. He was promptly jeered by the crowd and jostled off the sidewalk and into the street. At that point he produced a pistol, threatening to shoot the dogs. Members of both companies stepped in to halt the fight.

Rolla was gingerly picked up by his mates, placed on the engine, and taken to the firehouse. There they bathed him and tended his wounds. After a short while he walked outside and took up position where he could bask in the afternoon sun. He was none the worse for the experience and was ready to answer the next fire call. Eagle's hound never did bother Rolla again.

His devotion to duty became the center of many stories told around the enginehouse. One

particular favorite tale concerned the events of April 10, 1852. It was the anniversary of the great fire, which was commemorated each year by the ringing of the fire bell at high noon. Hearing the alarm, Rolla left his master's Smithfield Street home and went to Third Street. He peered down Third looking for Vigilant's machine but there was no engine in sight. Puzzled, he looked for the Duquesne brigade but, again, no firemen were to be seen. As he was trying to figure out all of this, a gang of boys ran down Second Street pulling a handmade wooden fire engine. Rolla decided that any fire engine, even a toy one, was better than none and trotted along with them, joining their "company."

Alas, Rolla met his end in 1852 at the hands of some scallywag who poisoned him. His death brought great sorrow to Fulton and the members of Vigilant. He typified the firehouse mascot, ever faithful to fireman and fire duty. On the company rolls he would be remembered as a true and loyal friend of the Vigilant.

On July 3, 1852, Vigilant left Pittsburgh for an excursion to Wellsburg, West Virginia. The following account of the trip was reported in the pages of the Pittsburgh Leader newspaper upon their return to the city:

"On July 3, 1852, with red shirt, belt and glazed caps, with the word Vigilant in gold on the front of the same, over 50 equipped members left the engine house with the engine and hose carriage, and embarked on board a steamboat for Wellsburg, West Virginia, where they had been invited by the Enterprise Fire Company of that place."

The boat started and all was fun and frolic. Arriving in Wellsburg in good time the next morning, the Vigilants were welcomed by Dr. J. N. White on behalf of the Enterprise, and he was replied to by one of Mr. James Petrie's unexceptionable speeches, on behalf of the Vigilants.

In turn, the Wellsburg newspapers reported this a few days after the Vigilants returned to Pittsburgh . . . . . "It gives us an opportunity of

Old St. Paul's Cathedral, standing on an elevated island of earth, was consumed by fire in May 1851. The old building was left stranded by one of the early series of excavations to eliminate the "Hump." (Carnegie Library of Pittsburgh)

No matter how close the Reds and Greens came together in the cramped space, they did get along in splendid fashion.

doing honor to the zeal, efficiency and gentlemanly conduct of our guests, the Vigilant Fire Company of Pittsburgh. The Vigilant, with Byerly's Band, to the number of sixty-three, arrived early Friday morning on the board the steamer *Excel*. The 10 a.m. procession consisting of the Vigilant Company of Pittsburgh, the Enterprise, Union and Huron of Wellsburg with invited guests formed in front of the fire house. In the evening there was a torchlight parade and a grand fireman's ball. The next morning each company played their engines and all finally assembled in front of the Exchange and with several speeches bid farewell."

"The Vigilant was then escorted to the *Heroine*, which lay at the wharf waiting to convey them to their homes. Throughout the whole proceedings the utmost good feeling prevailed and, non withstanding the crowd, not a drunken man was to be seen nor did we hear a single oath. Our guests left behind them, and we are sure took with them, only feelings pleasant and fraternal. The company left the wharf at eleven o'clock, amid the cheers of the citizens of Wellsburg and the waving of handkerchiefs by the fair Virginians."

Arriving at Steubenville, the engine was unloaded and paraded through the main streets of the town. The trip was typical of the times. Firemen from one city often visit another, especially for a holiday parade or some similar celebration. This is where the term "visiting firemen" probably originated.

The year 1853 was the second in a row that started with severe winter weather. Local firemen were kept busy fighting not only fire but mother nature's elements as well. The year was even more stormy for the Niagara Company, which temporarily ceased operating due to internal struggles. Management and the membership were at odds over day-to-day operations of the company and when the problems could not be resolved, the brigade closed its doors.

Neptune finally realized a long-anticipated goal with the construction of a new and more modern enginehouse on Seventh Street. By January 1854, the Blues were able to move their engine in. Dedication for the new facility was held on August 19th. Eagle also built a new house to replace the one destroyed by fire in 1851. The new location was on Fourth Street near the intersection with Liberty. Money to build the new facility came from Eagle's treasury and subscriptions from local patrons.

This year was also remembered for the cholera epidemic that struck the city. Among the fatalities were several volunteer firemen, from Eagle Company in particular, notably Henry Plants, John Buchker, J. M. White, J. N. Perchment, and A. Roderbauch.

Many large fires challenged the local volunteers during the year 1854. Seventeen homes in Allegheny City fell victim to arson on January 7th. Three weeks later, Allegheny was the location of the Presbyterian Theological Seminary fire on Monument Hill. After those two spectacular North Side blazes, fire visited south of the city, first at Ihmsen's Glass House on March 22nd and then, the largest in that area, another glass factory owned by Johnson & Co. Before it could be brought under control, the May 22nd fire spread to 49 other buildings, reducing them to ashes. In downtown Pittsburgh, chemicals at the Fahnestock Drug Co. provided some powerful pyrotechnics when ignited by an October 2nd fire.

Fighting fires and just plain old fighting were staples in the daily life of local firefighting volunteers. The brethren were always ready to help each other in time of need or crisis, but flip the coin and their competitive, combative, jealous side would appear. Vigilant experienced more than its share of this trouble in 1855. Both human suffering and apparatus damage were incurred, on more than one occasion. During one attack, bricks were thrown, compelling the red shirts to abandon their engine to avoid bodily harm. Upon returning later to retrieve it, Vigy's squad found their engine bent and broken.

Along with their daily routines, volunteer firemen were always experimenting with new ways to make firefighting more efficient. With each alarm answered, some new idea or technique was introduced to try and improve the fire service. One of the problems that had long existed was how to replace the manual, hand-powered method of pumping up water from the engine with some sort of mechanized system. Several practical ideas centered on generating steam to power a mechanical pump.

Experimental designs evolved through 1854 and 1855, but without any positive results. Trial and error seemed to be the favored method of construction, with each new fire being a working test of the previous week's efforts. But persistence paid off the following year. One of Neptune's members was Joseph L. Lowry, who at one time was employed as an engineer in the municipal water system. His wide experience in water transmission and hydraulics was considerable, and it was this expertise that led him to build Pittsburgh's first steam-powered pumping fire engine, the "Citizen."

When completed, it was put in the care of Neptune for their use and further testing. Compared in size to the other city engines it was a giant. Controls and gauges were set out on a side control panel. A large vertical boiler housed a copper coil, which was heated by a firebox at the base. Water to make steam was stored in a cylindrical tank that sat forward, interconnected by a network of pipes and valves. The entire machine was trimmed in nickel and brass. A painted carriage with steerable tongue carried the assembly. Flying from a mast at the engine front was a blue flag with a white stripe that bore the company motto: "Beat this engine and take the banner." During tests, water pressure was dramatically increased, as was the length of the stream projected. Only two shortcomings were noted. It took far too much time to build a fire to heat the boiler water and thus make steam. And second, the engine maneuverability was poor due to the heavy chassis loads. In time, the experimental engine was retired from service and dismantled.

A lighter weight engine was then ordered from William M. Jones & Co. of Baltimore, Maryland. Also steam-powered and constructed to beautiful lines, it, unfortunately, was not built as sturdy as the local machines and literally fell apart while working. Neptune's management was unhappy about this to the tune of $2,700 for repairs, part of which was shared by the city. No matter, steam power had proven its value and would be the basis of all future fire engine designs. Local volunteers no doubt wished they had at least one or two at the Phillips-Best Flint Glass Works fire during 1856. Insufficient water delivery by volume and distance caused not only the loss of the manufactory, but also 50 houses surrounding the Try Street site as well.

And while some of the Pittsburgh volunteer fire companies were enjoying a measure of prosperity, others were plagued by everyday misfortunes or fiscal difficulties. One of these outfits was the Relief Fire Engine Company, chartered shortly after 1850. Money and the search for it seemed to be at the center of all company business. The official minute book entry of October 2, 1855, stated that "A comity was appointed to solicit superscriptions for the benefit of the company." December 28th's entry read "Resolved, that provided the money is in the treasury, the President shall draw his warrant in favor of D. Littel for $25 for service about the engine and hous."

Fortunes looked brighter in 1857 when a February 7th resolution requested that "a committee be appointed to get 50 badges, the badges to be blue with gilded letters." Company spirits were also much improved when, in concert with Niagara, Vigilant, and Good Intent, Relief petitioned city councils for $250 each for the purpose of buying new hose carriages.

In the meantime, it was back to the drawing table, forge and tool chest in the quest for a mechanized engine. To build a successful steam-powered fire engine, several important elements were needed. First was the capability to quickly produce steam to the pump mechanism in no more than five minutes, and second, the entire rig must be easily transported through the congested streets of the city and maneuvered into tight locations where necessary. The result should be a stream of water traveling no less than 150 feet from nozzles end to the fire.

After many more months of trial and error, and equally many months of dismal performances or outright failure, a successful design was built by Eagle Company. It was fitting that the first organized volunteer fire company should also be first in this endeavor. But victory did not come easily. The machine was a totally home-grown effort by the members of Eagle that quickly became the subject of skepticism and ridicule by the other volunteer companies.

Eagle Company's pride led it to take the first important steps to this end in October 1859. A five-member committee consisting of Thomas Rees, Henry Moreland, Columbus West, George Wilson, and Alexander Gracie met with draftsman James Nelson to discuss the design and plan the construction drawings. James Rees' forge produced all the castings for the engine, and assembly of the components was started under the supervision of George Wilson. Upon completion, it was decided to test the engine in secret, at night, just in case of mechanical complications. The first tryout was held at midnight on Water Street and, as feared, the engine refused to pump.

Members of the opposition, gathered nearby and out of sight, let out with catcalls and a cackle of mocking laughter. One detractor called out to Eagle's builders: "Take your old steamboat home, she can't squirt!" Unruffled by it all, Eagle's committee collected its equipment and took their engine back home to fix the problem. In a short time the trouble was traced to a leak on the vacuum side of the pump and repairs were made. At the next test the engine performed flawlessly, silencing all critics. Pittsburgh's first steam-powered fire engine was now in service thanks to the men of Eagle. It would be the first of a long line of modern engines to serve the city.

An early commercially built steam fire engine delivered to Pittsburgh was this Amoskeag unit ordered by the Neptune Company in 1862. (Carnegie Library of Pittsburgh)

And while some of the Pittsburgh volunteer fire companies were enjoying a measure of prosperity, others were plagued by everyday misfortunes or fiscal difficulties.

Niagara Company also designed and built a steam-powered machine in 1859. It was fabricated locally at the Knapp Machine Works, with members James Hemphill, Joseph French, Joseph Kane, and John McElroy providing the necessary expertise and assisting with the labor. A total bill of $2,300 for materials and labor was paid by commercial establishments around Niagara's headquarters plus several local property insurance underwriters.

Progress was also evident in Allegheny City, which by 1860 boasted four major volunteer fire brigades. These included upgraded volunteer companies William Penn and the President, plus the newer Uncle Sam, and the modernized Phoenix Engine Company. Added to the rolls this year was the recently reorganized Hope Engine Company. City government also initiated the first steps toward what would eventually become a paid employee fire department. Legislation was enacted to allot each volunteer fire company Captain living space within the enginehouse and pay him a salary of $50 per year. Each company was also permitted to choose four members, each of who, because of their close proximity to the enginehouse, was paid the sum of $40 each, per year, plus $1 each for every fire call answered. These firemen were designated as Chief Volunteers.

Relief, too, was in the hunt for steam-powered equipment. An executive committee meeting was held in February 1861 with the purpose of establishing a committee to purchase a steamer. Those honored with the task included: E. S. Wright, B. C. Sawyer, J. E. Swint, George Cochran, Jr., and Christopher Oyer. The search was interrupted by the war but resumed in December 1863. Official action then led to the renaming of the company to the Relief Steam Engine Company. It is uncertain if indeed a steam-powered fire engine was ever procured by Relief. Unfortunately, the company had earned the reputation of a social club rather than a volunteer fire company dedicated to the fire call.

Vigilant, meanwhile, had spent considerable time and effort gathering information on steam powered firefighting machines. Seeing a chance to watch one in action, its members asked Allegheny's Hope Steam Engine Company to put on a demonstration upon their return from a trip to Chicago. Hope agreed, and put the engine through its paces to the delight of the red shirts and various onlookers. Vigy was sold on the idea and proceeded to arrange the necessary financing to purchase one.

With about $2,000 in their coffers, Vigilant organized a committee to go forward and make the purchase. In October 1860, James Irvin, Eugene Alexander, B. C. Sawyer, Sr., Robert C. Elliott, and James Petrie headed to eastern Pennsylvania and the major east coast cities to visit the factories of all the steam powered fire engine manufacturers. It was a whirlwind tour, with each company trying to outdo the other in order to get a signed contract. In the end, the Manchester, New Hampshire Locomotive Works was selected to build a second-class Amoskeag-type engine and deliver it to Vigilant by July 1861.

When the new engine arrived that summer of 1861, it was a landmark day for Vigilant and the city. This was the first commercially built steam-powered fire engine placed in service in Pittsburgh. A working demonstration showed that the polished brass and brightly painted machine boasted more than esthetic beauty. Not only did it project a stream of water almost 200 feet, but it also sustained that play for five hours straight.

The once proud "Red Bird" engine, John Agnew's handiwork, sat quietly at the rear of the engine house, Its years of faithful service were at an end. Replaced by the modern miracle of steam, the old veteran was sold for $1,450 to the town of Salem, Ohio.

Vigilant moved quickly to enhance its reputation as one of Pittsburgh's premier volunteer fire companies. New quarters at Third Street, near Market, were purchased and extensively renovated. Member Robert C. Elliott was in charge of the work. Many of the improvements, such as indoor plumbing, were considered lavish for their day. Local newsprint named it the "Fireman's Palace," and called Vigilant itself the "Pride of the West." No matter, Vigy had its own idea about names, and in November 1861 changed its official company title to Vigilant Steam Fire Engine Company and Hose Company.

Having the new equipment would only help to maintain the status quo in protection, as major fires

Vigilant's Third Street enginehouse was a modern brick veneer on wood frame affair, with living accommodations on the second floor. (Collection of Richard L. Linder)

continued to occur. On September 17, 1862, a mammoth explosion leveled the Allegheny Arsenal Laboratory, touching off a fire peppered with exploding ammunition and ordinance. The year 1863 marked the loss of the Third Presbyterian Church at the northeast corner of Ferry and Third Streets. These were major fire events in which the structures were a total loss. It remained for the volunteers to confine the fire and keep it from spreading. To have any chance at all of saving buildings that were on fire, additional men and equipment would be required. It was questionable whether the present volunteer system could provide adequate protection for all the people of Pittsburgh.

Allegheny City management decided the time was right to get involved in the quest for a steam-powered engine. Much like their cousins across river, investigation, building, testing, and trial and error were the normal course of things. The result was the purchase of a steamer on July 23, 1863, which was christened "Old Hope." It was delivered to the town fathers and was turned over to the care of what was now the Hope Steam Engine Company on August 1st of that year. A second new steam engine was assigned to the General Grant brigade on November 5, 1863.

Another type of battle was being waged by the Niagara Company volunteers about this time. The American Civil War was well under way and one of the Union's more noteworthy fighting units was the 102nd Pennsylvania Volunteers, made up almost entirely of Niagara firemen and one fire dog. That canine was "Jack," a white and gray bull terrier that held the distinction of being listed in the war's official records. Years earlier he had come to the Niagara Company in a most interesting manner. Not the shy, retiring type, he decided to get personal with a mule on Penn Avenue, across the way from the Niagara enginehouse. Taking exception to the terrier, the mule used its hind quarters to promptly dispatch Jack through the air, depositing him in the Niagara's basement.

It was there that fireman Hickory Jones found Jack, unable to move because of a broken leg. Jones gently carried him to the first floor and proceeded to tend his injury, using splints cut from a wooden shingle. Jack convalesced at the enginehouse, becoming attached to the volunteers, and they to him. Soon he was up and around, responding to alarms along with his brigade and its engine. His service was as faithful whether it be day or night, sun, rain, or snow.

In time, the call to arms came, and the men of Niagara departed for the war. So did Jack. While on duty, he proved himself to be an excellent picket guard. Off duty, it was said he slept with one eye open and both ears perked. He actually participated in the charge with his regiment at the battles of Fair Oaks and Williamsburg, Virginia, and was wounded at the skirmish on Malvern Hill. Later he crossed the Rappahannock with the 102nd to fight both of the battles of Fredericksburg.

At the battle for Salem Church, Jack was taken prisoner and later exchanged for a prisoner from the opposing side. On the 102nd's return through Frederick, Maryland, Jack disappeared, never to be seen again. His fate was unknown. The hearts of the Niagara Company were heavy upon their return to Pittsburgh. Jack's loyal presence was sorely missed and after their return to Pittsburgh, they kept a photograph of him, along with a copy of his war record on the enginehouse wall.

Progress continued in the quest for newer and more modern equipment, as evidenced by Niagara's purchase of a new Amoskeag steam powered single-pump engine in August 1865. Replacing equipment was a slow and painful process. The biggest question was not always the what, but more likely the "how." Everyone, it seemed, wanted progress, but nobody wanted to pay for it. Thus began the slow and painful process of getting subscriptions, sponsors, or help from the city treasury to fund the purchase.

By the late 1860s, Allegheny City had surpassed Pittsburgh in the number of companies, with the addition of the Columbia, Ellsworth, and Friendship Volunteer Fire companies. A network of fire alarm boxes was also built in mid-1867 to serve the main area of the town, giving ready access for activating the fire call, thus saving precious time. Allegheny's first fire service supply and fuel wagon was placed in service during 1869.

History also records the names of other volunteer fire organizations active around the three rivers during the period of the 1820s to the 1860s. Some were well established groups with men and equipment, some were little more than a name on paper. Their names are worth remembering, if not for their work, then for their noble intention to serve. They were the Ben Franklin, Rescue, Good Will, Blue Dick Hose Company, Fairmount, and the Humane Hose Company.

The end of the decade proved, if nothing else, that firefighting was becoming a full-time occupation. There were more and more fires being fought with the same or less manpower. Some were poorly trained for the task. Advances had been made in equipment design and performance, but more full-time firemen, skilled in the art of fire work, were needed. In truth, over the past 70 years, the volunteer fireman's job had changed little in comparison to the circumstances and environment he was called upon to work in. It was a situation that demanded a solution, and quickly. The answer would bring big changes to the municipal fire fighting system and its members. The days of the old volunteer firefighter, his oilskin coat, and pointed metal hat were numbered. A new era of firefighting was about to dawn. ☆

The end of the decade proved, if nothing else, that firefighting was becoming a full-time occupation.

Chapter 3

Getting Up Steam
1870-1899

Paid professional firemen take the reins; the volunteers answer their last alarm.

As a new year and a new decade started, one legendary volume of Pittsburgh firefighting history was about to come to an end and another epic chapter was ready to begin. For some time now, there had been growing concern by the city fathers that the present system of fighting fires with volunteer manpower was a method obsolete, if not downright dangerous. While Pittsburgh had prospered and expanded over the years, the methods used to fight city fires had changed little in 75 years. The core of the system was basically a group of individual companies that would answer fire alarms, providing that enough men showed up to pull the engine and lay hose.

In the later 1860s, Pittsburgh's councils, over the opposing petitions of several local volunteer fire groups, officially requested that the state legislature amend the city charter language to authorize the implementation of a full-time paid fire department. Without debate, the necessary legislative bills passed through the Pennsylvania general assembly and were signed into law by Governor John W. Geary on March 23, 1870. The city of Pittsburgh now had the right, as official language put it, "To establish, organize and control a paid fire department in and for the City of Pittsburgh to and provide for the expenses thereof." This money was to be raised through a 3-1/2 mill tax on the gross dollar profits of all insurance companies doing business in the city.

Members of council then started the arduous task of organizing the newly created Pittsburgh Fire Department. They were assisted in their task by Charles T. Holloway, Fire Inspector and former Chief of the Baltimore, Maryland Fire Department. Holloway was also engaged in the manufacture of fire fighting equipment and had extensive experience in fire department matters. He held the distinction of assuming the presidency of Baltimore's Hope Junior Fire Company at the age of 15. The end result was a new city ordinance enacted on April 14, 1870. Among other language in the act was that which gave fire department employees and vehicles the right of way over any other except those transporting or delivering the U. S. Mail. A board of nine commissioners was appointed to convert the existing volunteer system to a paid municipal employee department and manage it on a day-to-day basis. Those installed as commissioners were industrialist M. K. Moorhead, Robert Finney, R. W. Mackey, financier Henry Hays, merchant Thomas Reese, W. M. McKelvey, John H. Stewart, businessman John H. McElroy, and John J. Torley, himself a volunteer fireman.

On May 15, 1870, the Board of Fire Commissioners held its first official meeting and elected Henry Hays president and John McElroy chief engineer. In other non-board appointments, W. B. Neeper was elected secretary; William White, assistant chief engineer; and S. T. Paisley, superintendent of the Fire Alarm Telegraph. It was their sworn duty to uphold the city ordinance, which read, in part, "To form a fire department, said department shall have sole and exclusive power and authority to extinguish fires in said City of Pittsburgh." Yearly wages for line officers and other employees were set at $1,200 for the chief engineer, $840 for the assistant chief engineer, $840 for each foremen, $820 for each steam engine engineer, and $720 for each engine driver. All other firemen were paid a flat rate of $720 per year.

Planning and forethought was evident in the designation of a position of engine driver. To date, practically all engines were pulled by ropes to the fire, using human engines. On occasion, horses were used as prime movers, especially at parades, where they were dressed up and gaily decorated to pull the company engine. But, in most cases, manpower did the job, and had done so since the very beginning. Now the lighter-weight hand-operated pumpers were giving way to steam power. The

opposite page, **First National Bank Building after the June 10, 1886, fire. The extension ladder truck, with twisted upper structure, is set up on Fifth Avenue. Wood Street crosses, left to right, near the bottom of the photo.** (Carnegie Library of Pittsburgh)

Act Establishing Department. 87

AN ACT
TO
ESTABLISH THE PITTSBURGH FIRE DEPARTMENT.

SECTION 1. *Be it enacted by the Senate and House of Representatives of the Commonwealth of Pennsylvania in General Assembly met, and it is hereby enacted by the authority of the same,* That the Councils of the City of Pittsburgh be, and they are hereby empowered to establish, organize and control a paid Fire Department in and for the City of Pittsburgh, and to provide for the expenses thereof.

SECTION 2. That for the purpose aforesaid, and in addition to other revenues, it shall be lawful for the said Councils to assess and collect a special tax upon all Fire, Marine and Life Insurance Companies or Agencies doing business within said city: said tax shall not exceed in any one year the amount of three and one-half mills on the dollar of the gross receipts of said companies.

SECTION 3. That said Councils shall have full power to make all necessary and proper Rules and Ordinances for the government of said Fire Department, not inconsistent with the constitution and laws of the State of Pennsylvania, and to enforce the same by proper penalties.

BUTLER B. STRANG,
Speaker of the House of Representatives.

CHARLES H. STINSOM,
Speaker of the Senate.

Approved, the twenty-third day of March, Anno Domini one thousand eight hundred and seventy.

JOHN W. GEARY.

The official instrument that authorized the City of Pittsburgh to establish the paid Pittsburgh Fire Department.
(Collection of Richard L. Linder)

John H. McIlroy was appointed chief of the newly formed Pittsburgh Fire Department in June 1870. (Collection of Richard L. Linder)

new "kid" was much heavier and harder to maneuver. In the old days, stopping the wooden machine always meant facing the risk of being run over by it. This new invention, larger, with metal boiler and on-board water tank, tripled those dangers; something new and safer was needed.

Pittsburgh Fire Department Commissioners decided to phase out the manual pulling of engines by the rank and file in favor of horse-drawn power for all engines of the department. The average price paid for a department horse ranged from $165 to $225 each. Thus, the title of engine driver or driver of steamers was added to the official firefighters vocabulary. Three basic types of firefighting companies were set up by John McElroy and William White. Steam engine pumper companies were subsidized $3,000 by the city and were made up of eight men: captain or foreman, engineer, fireman, driver, and four hosemen. Next, each hose reel company was funded $1,000 by the city and consisted of four men: foreman and driver plus three hosemen. Finally, companies with ladders and hooks were manned by six firemen; one foreman, one driver, one tillerman, and three ladder men.

Work began at once to inventory all usable pieces of equipment owned by the individual volunteer fire companies. From this list, a determination was made concerning which equipment would be kept and reused in the new department and what would not be suitable or needed. Each existing brigade was payed an appraised sum of money for items kept by the city for inclusion in the new department. Eagle's steam engine was purchased by the city for $3,482. On June 6, 1870, the city closed the sale of Duquesne's engine for $3,800. Seven days later, on the 13th, the city paid $7,940.31 to the Independence Company for their equipment. After reaching the final settlement with the city for all of its property, the Duquesne brigade members split the monies equally between them, each one receiving about $150. Any unwanted equipment was returned to its owner. By years end the newly organized Pittsburgh Fire Department was in place and functioning. Employee salaries and operating equipment totaled $66,555.77 for the period.

It did not take long for the newly constituted fire companies to meet their foe head-on in a major event. Crude oil stored in the Standard Oil Company depot along the north bank of the Allegheny River erupted in fire during October 1870. Firemen answering the first alarms from nearby telegraph box No. 85 were greeted by fire that was traveling on a burning sea of flowing oil, rolling out to meet them from broken and burning barrels.

Standard's watchman perished in the blaze, roasted to death by the petroleum-fed fire, which melted the firemen's gum slickers and singed their eyebrows and hair. In order to breathe, firemen wet down their handlebar mustaches and long beards and placed them in their mouths as filters so they could inhale without taking in superheated air. Human eyeballs felt like they were roasting in their sockets; nostrils were scorched by flames and fumes from the gigantic fire. In some places, the flaming oil was knee-deep, running downgrade to the river. At the shore, the flames found more fuel in the loaded barges used to store large quantities of oil for transport. They burst into a giant fireball that pushed back the attacking hosemen with a wall of blistering heat and oily black smoke.

It was the worst fire in Allegheny City to that date. Two bridges, several boats, and assorted riverfront properties were consumed by the fluid path of the fire. The river surface itself was on fire for three days. For the Pittsburgh Fire Department, it was truly a trial by fire.

In January 1871, the fledgling Pittsburgh Fire Department employed 69 men. A combined equipment roster listed six steam powered fire engines, no two of which were of the same manufacture or identical design. Three engines were second-class, double-plunger types, the remaining were second-class harp design. Assigned for duty with each engine company was a hose company. There were two hook and ladder carriages, each carrying assorted ladders, totaling around 150 feet in length, plus a variety of hooks, ropes, buckets, and poles. Held in reserve was one steam engine company with hose brigade. In support, if required for any particular alarm in any part of town, were two hose reel companies, one located near Soho Run and the other in the 17th ward. One engine was not in service due to its heavy weight bearing on antiquated wheels and axles. Twenty-four horses shared pulling duties at the equipment drawbars. There was a total of 97 fire alarm telegraph signal boxes interconnected by 65 miles of wiring.

With the arrival of the new year of 1871 came the realization that more than the old year had been left behind. There was a mixture of sadness and uncertainty shared by all city firemen. The volunteer fire company, of which all the 'new' firemen were once members, and a civic fixture for more than 75 years, had passed into history. All of the sweat and toil at the pump handles, in all kinds of weather, the suffering and sometimes death shared by comrades of all the brigades, all of these things which welded them to the common cause no matter what individual company loyalties dictated, was gone, part of a glorious, proud and dedicated heritage. Who could predict what would lie ahead? Would the new experiment work? Only time would give the answer.

Change was not only swift but impersonal as well, with no regard for sentimentality or tradition. Colorful engine company names like Eagle, Duquesne, and Neptune were replaced with common engine company numbers 1, 2 and 3; Relief, Independence and Niagara companies became numbers 4, 7, and 15 respectively. At least some of the old volunteer companies survived the reorganization. For some brigades it was the end of the line. The final bell had been answered. More than one half of the original companies disappeared from the rolls, among them the Tans of Allegheny. At the stroke of a brush or pen, seven decades of volunteer fire department service were erased and a new era was at hand.

Firehouse life was now a new and different experience. Personnel were initially on duty during daylight hours and on call at all other times. All of the elderly hand-pump engines were retired, either sold to little communities that needed a small-capacity engine, or rolled into storage at an out-of-the-way garage or barn. For most of the firemen, it was like beginning again, getting acquainted with the new steam-powered wonder. But they were a resilient bunch and quickly, through trial, error, and on-the-job experience at fires, mastered the fundamentals demanded by this latest technology. Not only was there a call to duty on the land but on the water as well. Several companies pulled up for hose work on the Monongahela wharf August 14th to extinguish the burning hulk and scattered framework of the steamboat *Chautauqua* whose boiler had exploded without warning. Eight rivermen were killed by the thundering blast, which also shattered windows in buildings along Water Street.

Most of the newly constituted fire companies exhibited the old pride in caring for their enginehouse and equipment. Elbow grease plus spit and polish was the order of the day after returning from an alarm. A clean engine seemed to work more easily, travel faster and certainly looked better. The steam engineer would clean out his firebox and set fresh kindling, ready to be lit at the next alarm bell. Getting steam up quickly was almost as important as rushing to the scene of a fire itself. And to get man and machine to the blaze, firemen relied on their horses. From this reliance a new bond of tradition was founded. Firemen and firehorses, along with their firedog mascots, became an inseparable team, working together to serve the public.

There was trust, loyalty, and even affection between most fire companies and their working steeds. Some horses were known to kick their stall boards to help keep the night watchman

Firemen and firehorses, along with their firedog mascots, became an inseparable team, working together to serve the public.

These horses were the first team in the Pittsburgh Fire Department to be outfitted with the new quick adjustment harnesses. The young boy to the right of the horses, in dark clothes with his hand on his hip, is Thomas L. Pfarr, later the city fire marshall.
(Carnegie Library of Pittsburgh)

awake. Many exhibited gentle personalities as counterpoint to great physical prowess under the harness. Two beautiful matching chestnut browns belonging to No. 3 company were widely regarded for their friendliness with station visitors and their love of racing to fires pulling their engine. The four grays of Number 6, Sam, Dan, Bill and Jim, were a sight to behold, turning a corner in tandem or running flat-out on the straight away. Quick response to the fire gong was always imperative. It took precious time to bring up the horse team and hitch them to the engine tongue. Experiments were tried to see if time could be reduced from initial bell to actually exiting the enginehouse. Local talent provided the first practical method of rapidly hitching up a double team.

A device called the "quickhitch" was invented by firemen in Allegheny City in 1871. Essentially it was a pivot-jointed harness attached to the engine frame and tongue, held overhead until the horses were in position, then lowered, attached to the animals and disconnected from the suspension mechanism. In minutes, a company was out the door and on its way to the fire. This first successful hitch device cut starting time by half and enabled a fire company to be on the scene and working quickly.

With the appointment of Andrew Jackson Cupples as chief engineer of the department, the rank and file firemen had a Champion, sympathetic to the firemen's lot in life. In support of a request to raise the basic pay for city firemen, he wrote, in part, " Only a man who is actuated by such a high sense of honor to cause him to perform well and faithfully his duty in whatever situation in life he may be placed, is capable of performing the arduous duties of the fireman, to take the risk of both life and health which so often falls to his lot. Many such men, I am proud to say, you have in this department." The joint city councils must have agreed with him because they passed a resolution on May 29, 1871, to increase the salaries of all firemen from the lowest rank up through Cupples himself.

Along with the establishment of a new municipal fire department, ground was broken for a new city water supply and distribution system in 1871. This was the first building block in the greater Pittsburgh water system that would be expanded in the coming years to meet, among

Ready for parade duty, an unidentified Pittsburgh Fire Department unit turns out in suits and straw hats with canvas engine covers in place. (Carnegie Library of Pittsburgh)

other needs, increased water gallonage for fighting fires from the city's 410 water plugs. At the heart of the works was the newly designed Brilliant Pumping Station, located on a tract of land six miles up the Allegheny River from the city.

Insurance losses due to fires fell again to $146,482, of which half of this amount was attributed to one fire. Once again, the volunteer spirit appeared briefly during November 1871. At the beginning of the second week of the month, a disabling horse virus put all the four-legged fire company members out of action until mid-December. In order to get the engines and other apparatus to the fires, ordinary citizens took up the tow ropes and did the work. No shortage of manpower was noted.

There was also a pressing need for new fire department equipment, a fact that the Board of Fire Commissioners presented to the city councils. By January 1872, construction was underway to build a new engine house in the city's Hill District. On April 4, 1872, Engine Company No. 5 was activated in a new building at 2155 Center Avenue. Their activation came none too soon, as the blaze at the Fahnestock Lead Works on Liberty near 20th Street required a maximum effort to put out.

Engine repairs were a constant, almost everyday occurrence. Many engine companies kept spare parts on the premises, but major repairs were accomplished by outside concerns. Additional apparatus was needed to upgrade the older equipment and add engines to cover the expanding city neighborhoods. The tax imposed on insurance companies doing business in the city failed to bring in sufficient revenue to offset fire department expenditures. Finally, in 1873, at the behest of Chief Engineer William J. White, the City of Pittsburgh borrowed $200,000 for the express purpose of buying new fire engines and upgrading the fire alarm telegraph box system to cover the boroughs south of the Monongahela River. These were annexed to the city the previous year.

During 1873, major fires, a few months apart, tested the mettle of No. 5's enginemen. Work began on May 9th at the Pittsburgh Novelty Works, located at Grant and First. The structures were three two-story brick affairs that had fallen into a state of disrepair. Pittsburgh Novelty used the building as a warehouse for storage of its cast-iron products. Upon arrival at the scene, company hosemen entered the burning structures to try and locate the source of the fire. Orders soon came to pull back and the firemen did it none too soon. A minute after they evacuated, the burning buildings collapsed. Later investigations fixed the cause of the fire as carelessness. The amount of loss was $47,864, of which insurance covered a little more than $35,000.

While fighting the Lewis Foundry fire on July 12, 1873, firemen held wooden shields between

them and the fire so they could get close enough to direct water on the flames. Even so, their hands and faces were blistered and their hair began to smoke. Within 30 minutes, the fire went to three alarms. On the Second Avenue side of the building, a man in a buggy started to drive past the building. A wall of searing heat hit his rig, knocking the horse down and sending the man stumbling along the ground to safety. He was severely burned, and his horse was reduced to charcoal. Two adjoining homes on Try Street were also consumed in the blaze.

With each day, each fire call answered, the paid city firemen became more efficient in their appointed tasks, more important as cogs in the overall Pittsburgh Fire Department operation. A decrease in the fire losses to $211,989 for the year was evidence of that. Signs of pride and competitive spirit, hallmarks of the volunteer days, began

Firemen had to contend with freezing temperatures and heavy ice as they fought this fire around the White Dental Parlors in Allegheny City. (Carnegie Library of Pittsburgh)

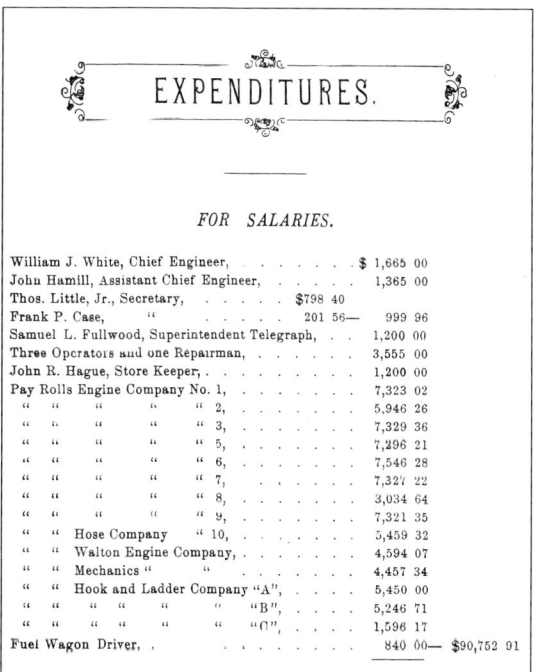

Fire department salaries for the year 1873 totaled $90,752.91. Curiously, fuel wagon drivers were paid more than the department secretary. (Collection of Richard L. Linder)

EXTRA FIRST SIZE DOUBLE STEAM FIRE ENGINES.
Crane-Necked Frame. Horse Draft.

Although this Extra First Size Amoskeag steam fire engine could be pulled by two horses, draft gear was available from the factory for a three-horse team. This engine could pump 1,000 gallons of water per minute. (Courtesy of the Manchester, NH, Historic Association)

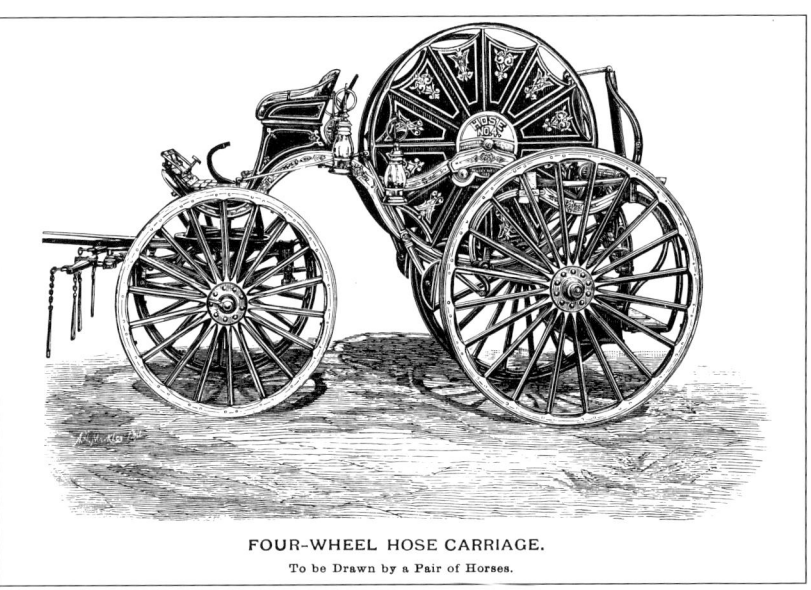

FOUR-WHEEL HOSE CARRIAGE.
To be Drawn by a Pair of Horses.

A Pittsburgh Fire Department favorite was this Amoskeag two-horse, four-wheel, hose carriage. Carrying as much as 1,800 feet of hose, it length measured 22 feet, 4 inches in length. (Courtesy of the Manchester, NH, Historic Association)

to reappear. Perhaps it was because all the new companies began service at the same time that made the individual groups want to compete with their fellow firemen for bragging rights. Or maybe it was because no one could claim that older in service was better, they all being the same age. Whatever, each company had to prove itself by individual work.

As a result, firemen themselves made personal contributions that improved daily operation of the firefighting service. One such innovator was fire fighter John J. Freyvogel of Engine Company No. 5. He devised a metal horse collar, which, when placed around a horses neck, adjusted to the neck contour for a perfect fit. The collar was hinged over the mane and connected underneath, much the same as a handcuff would do when applied around a persons wrist. Also, 1873 was the year that No. 4 Engine Company installed a mechanism that dropped horse stall gates, turned up gas lighting jets and swung open the station doors whenever the alarm bell sounded.

Statistics recorded during the year 1873 listed 10 steam powered pumping engines with hose carriages now performing duty, with an additional two engines and six hose reels held in reserve. Three hook-and-ladder trucks completed the lineup of major department apparatus. Newly activated engine companies included the No. 12 previously the old Walton Engine Company, located on Sarah Street between 20th and 21st streets on the South Side, and No. 11, formerly the Mechanics Engine Company, located on 14th Street between Washington and Carson Streets, South Side, Number 10 at 44th and Calvin streets, and the No. 8 located at Highland Avenue and Broad Street, East End. Manpower in the department now totaled 112 men. Anticipating the completion of the new, higher capacity water pumping station now under construction, new fire hydrants were installed in districts with little or marginal coverage, bringing the total to 663. The fire telegraph network also added new boxes in outlying neighborhoods to expedite the transmission of alarms into the city. The entire 27-square-mile city landscape was served by 128 alarm stations.

Engine Company No. 4, the old Relief Brigade, made its mark in Pittsburgh Fire Department history in May 1875. Company Engineer John Busha had noted that on some occasions there was a question or a possibility of a miscount in bell claps when the fire alarm sounded. The number of incoming rings denoted the zone or location of the fire. An alarm counted incorrectly, in the rush of getting mobilized, could lead to a serious loss of time or even tragedy. Busha studied the problem and came up with an attachment that bolted onto the bell tapper arm. This added piece, sort of an arm, held a stick of chalk at the other end that contacted a square of slate. When the alarm sounded, each bell strike in turn caused the chalk to mark the slate, giving a visual count of the strokes. It was a revolutionary idea, one that eventually was put into use by just about every city fire department in the United States.

By the halfway point of the 1870s, a new and different way of life had settled into city firehouses. All remaining vestiges of the volunteer enginehouse, especially in its role as a social center and community gathering place were gone. Now it was all business, serious business. The fire watch was kept 24 hours each day, with company personnel rotating through eight-hour shifts to man the alarm desk. It was the watchman's duty to note each and every alarm, marking it in the company log book. He also checked on the horses and fed them, and made sure that all was ready in case an alarm went off to call out their particular station. The remainder of the company was split into shifts guarantying that the station had a full crew around the clock. Station houses became the fireman's second home where he worked, ate, and slept but, most of all, kept ready to answer the fire bell. And there were plenty of alarms to respond to, a record 195 instances in 1875.

Soon, the department increased its size once again, this time to 13 stations, with the addition of Engine Company No. 10 taking up residence at Steuben and Mill Streets in June 1875. No. 10 had the distinction of having the only African-American firefighter employed by the department. He was engine driver Enos Hutchinson, who, in previous years, drove for No. 3 Company. Next came the No. 17 at Bailey Avenue and the Castle Shannon Incline in January 1877. Operational orders called for the company to answer alarms in the 21st, 27th, 32nd, and 35th wards.

Most of the streets in these wards were primitive, little more than dirt roads. And there were always hills to contend with. In winter, answering an alarm was an adventure. One snowy day, while trying to make a turn onto Virginia Avenue from Wyoming Street, No. 17's engine brakes failed. Firemen jumped off to save their lives. Fortunately, the leather harnesses were pulled apart, freeing the horse team and sparing it from certain death. The engine itself came to a crashing stop, sliding on its side into a neighborhood storeroom.

Five hook-and-ladder trucks were strategically located around town, ready to respond and assist with fires in elevated structures. Employee rolls for 1876 listed a total of 126 firemen of all ranks and titles, right up to Chief John H. McIlroy. Approximately 200 fire calls were answered during the year, a new record for the department. At that point, the paid city fire department had almost six years of experience under its belt. The experiment was an unqualified success, but firefighters experience and professionalism would be severely tested in the coming year. It was fortunate for the department that in May 1877, leadership for all city firemen was vested in Samuel N. Evans, chief, and John Steel, assistant chief.

Of all the major fires that Pittsburgh firemen would battle, the railroad riot fires of 1877 were, next to the great 1845 fire, the largest in total area burned. Not only were there the usual dangers connected with fighting the fire, but there also was the threat of violence at the hands of the

Men from Engine Companies 2, 19, and Truck Company "A" pause from their work to look into the camera's lens.
(Carnegie Library of Pittsburgh)

A contemporary period drawing of the Union Depot and Hotel on fire, a casualty of the 1877 railroad riots. When the strike ended, damages were set at $5 million for the four-day rampage.
(Carnegie Library of Pittsburgh)

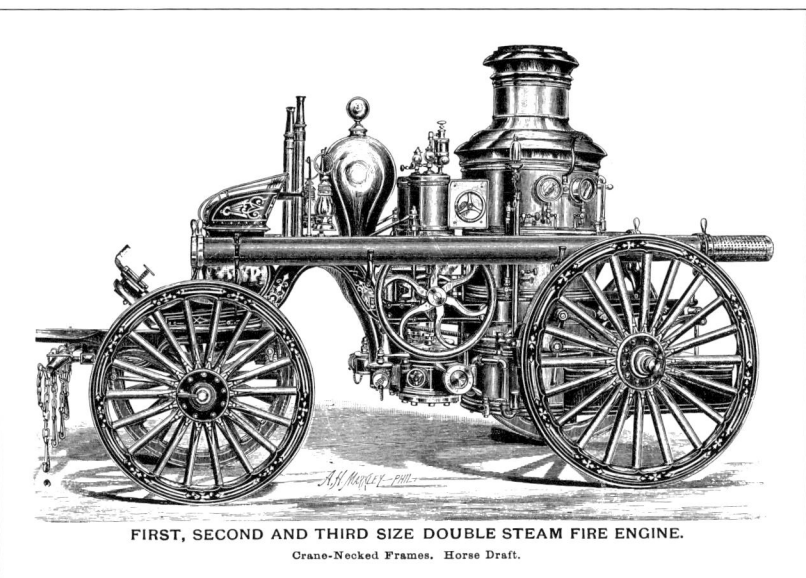
FIRST, SECOND AND THIRD SIZE DOUBLE STEAM FIRE ENGINE.
Crane-Necked Frames. Horse Draft.

Amoskeag's steam fire engines in the First, Second and Third sizes were similar in overall dimensions. They produced 900, 700, and 550 gallons of pumper water per minute respectively, which, when fitted with a 1 1/4 inch hose and nozzle, would deliver a stream of water 300 feet away. (Courtesy of the Manchester, NH, Historic Association)

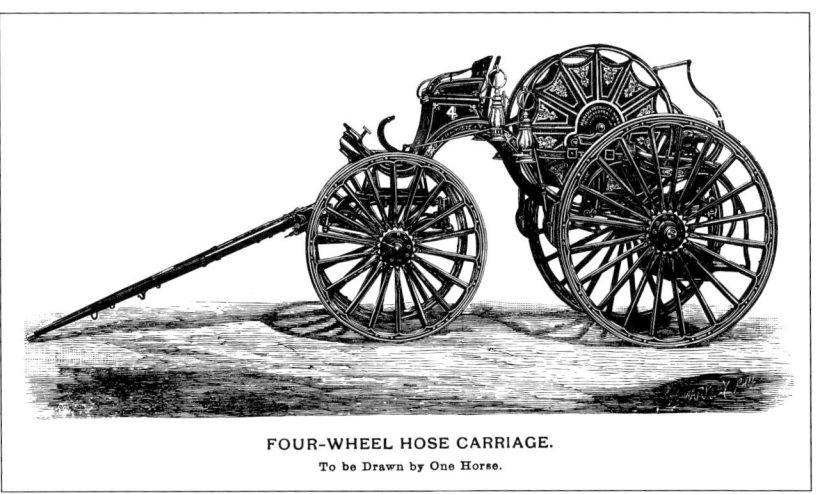
FOUR-WHEEL HOSE CARRIAGE.
To be Drawn by One Horse.

Amoskeag's single-horse, four-wheel hose carriage weighed in at 1,750 pounds unloaded, and carried 1,000 feet of hose. (Courtesy of the Manchester, NH, Historic Association)

> Of all the major fires that Pittsburgh firemen would battle, the railroad riot fires of 1877 were, next to the great 1845 fire, the largest in total area burned.

insurgents. Sparked by a 10 percent wage reduction for Pennsylvania Railroad trainmen, strikers refused to allow trains in or out of the company yards behind the Union Depot at Liberty, Grant and Penn. The situation deteriorated quickly and, when the state militiamen fired on the mob to defend themselves, total chaos broke out.

By nightfall on Saturday, July 21st, the Union Station itself was ablaze, plus the 28th street roundhouse and everything in between. Over 1,000 rail cars, 126 locomotives, and all company facilities were one giant mass of roaring flames. City fire companies responded to the multiple alarms shortly before midnight but rioters denied them access to fight the fire. Fistfights between firemen and strikers resulted in hoses being cut up, equipment being damaged and numerous personal injuries being inflicted. Firemen feared for their personal safety; some were threatened with death if they so much as lifted a finger to extinguish the burning railroad property. Company No. 15 was threatened with cannon fire and withdrew.

On the way out Penn Avenue to the blaze, the firemen of Engine Company No. 7 met an armed band of strikers who ordered them to turn back. No amount of reasoning would sway the rioters and No. 7 reluctantly pulled back rather than suffer damage to their equipment or harm to their person. Fearing trouble, Chief Evans ordered No. 7 to return to its enginehouse and stand by. When order was restored, the firemen returned to 26th Street, set up, and after bringing that section under control, proceeded to offer support at other locations where help was needed.

Members of the former Eagle Company, under the command of Thomas Cunningham, had a close brush with the rowdies on that Sunday afternoon. They were headed out Liberty Avenue to answer the alarm from Box 64 when the first drunken challenger, a large-framed laborer named 'Pat the Avenger," stepped out in front of their engine and bellowed: "Lookye here, if you fellows squirt one drop of water on railroad property you're liable to be shot and certain to have your hose cut." Cunningham alighted from his engine and came nose to nose with him and, poking his finger into the man's chest to make his point said: "Now you look here, if any man attempts to cut our hose he'll get his heart cut out!" That ended the confrontation and Eagle sped on to its duty station.

After much discussion and pleading by worried firefighters, strikers finally let them build a fire break to prevent the blaze from spreading into the adjoining, heavily populated neighborhoods. At Wylie and Washington streets, Engine Company No. 5 performed outstanding service in that regard, preventing fire from spreading up to the Hill district. Several times during the day, No. 5's hosemen had to turn their hoses on strikers to keep them from cutting their lines. Engine Company No. 4 put in 24 consecutive hours on the fire line, stationed at 25th and Liberty. The work was difficult in itself, spread out over many city blocks and complicated by scattered bands of rioters and sympathizers who took issue with anyone in authority like firemen. Things were finally brought under control on Monday, July 24th, but not before and estimated $5 million worth of mostly real estate and equipment of the Pennsylvania Railroad, had gone up in smoke.

The dollar losses, not to mention those who lost their lives in the insurrection, would have been many times greater, had it not been for the Pittsburgh Fire Department companies that successfully fought both flames and the unruly mob of strikers. Two of the many companies that served with distinction during this turbulent event were Engine Company No. 2, led by Foreman

Michael Hannigan and the No. 11 Company.

Meanwhile, in Chicago, Illinois, another firefighting tradition was born. On April 21, 1878, firemen of Engine Company No. 21 christened the first slide-down pole, a three inch round wooden affair, sanded smooth and oiled to speed firemen from the bunk room to their waiting engine. Two years later, the Worcester, Massachusetts, No. 1 Engine Company built and installed the first brass pole. In the Iron City, Engine Company No. 2 was the first Pittsburgh Fire Department Company to have a polished nickel-plated sliding pole, touted as one of the sights for visitors to see when touring the city.

Construction on the new, modern Brilliant pumping facility was completed in July 1878. Housed inside the plant were four mammoth engines driving dual pumps, which fed the new 280-million-gallon Highland Hill reservoir. Now the entire eastern section of the city would never want for water, especially at the fire plug. With Superintendent James L. Lowry in charge, plant operations started in the spring of 1879.

No. 11 company, along with Engine Company No. 2 and Hose Company No. 1, set a unique record in the autumn of 1878. They were selected by fire bureau management to quickly report to the nearby city of McKeesport, Pennsylvania, a result of an urgent plea for help in fighting a large fire that was threatening the community. Travel was scheduled over the Baltimore & Ohio Railroad main line between the cities. The companies departed Pittsburgh at 10 o'clock in the evening, loaded on a special train which had the right-of-way for the entire trip. Sixteen minutes later they pulled into the McKeesport station, having covered a distance of 14 miles on upgrades and curving track.

Another of Pittsburgh's contributions to the firefighters art occurred in the year 1880. Fireman John Freyvogel, of 'handcuff' horse-collar fame came up with another idea, which was quickly adopted and put into daily use. The device was operated by another new marvel, electricity, and at an alarm, it slapped the horses' flanks, signaling them to leave the station. Other firehouse refinements followed shortly. Now when the alarm bell sounded, horse stall doors opened and interior electric lighting came on. Some station houses installed outside double doors that would swing open at the push of a button when the company was alerted for duty. All of these things helped save precious time and get the firemen out of the enginehouse as quickly as possible.

Weather always played a large part in the work that firemen faced at the fire scene. Winter fires were particularly harsh and dangerous, adding another element of worry to the job. The Emory Oil Warehouse fire was one of those fires that was talked about for years after. On January 17, 1882, the temperature was near the zero mark. The cold was so severe that firemen took turns holding the nozzles, lest their flesh freeze to the brass. No sooner had No. 5 returned to its house, its firefighters covered from head to toe in ice, than another alarm struck. Off they went to 3rd and Duquesne Way, where 10 houses and a shop were burning. Firemen Hutchinson and Sparr had their lucky charms with them that day, both were trapped by a falling wall, and were rescued by their brother firemen.

As members of the Pittsburgh Fire Department went about their fire fighting duties, the variety of buildings they encountered were many and varied. Naturally, there were single family homes and tenements, easy prey for overheated cook stoves or carelessly tended oil lamps. Manufacturing companies and other industries were particularly vulnerable when forges or open flame were used in their work. Churches, schools, banks, and theaters were also regular victims. Every once in a great while, fire would break out in a landmark structure, one of particular or historical importance. That was the case on May 7, 1882.

It was shortly after the noon hour on that quiet Sunday when a Wylie Avenue pawn shop owner, Lewis Sussman, saw flames shooting out of the dome atop the Allegheny County Court House. In a panic, he ran down Fifth to Grant and, after securing a fire box key from the proprietor of the corner drug store there, activated the alarm at box number 26. Almost as fast as the fire telegraph made its report, word of mouth carried the news up and down the streets of the town, now filled with citizens just leaving church.

People hurried along the sidewalks, heading in the direction of the Court House, marked by a plume of grey smoke, rising into the clear spring sky. Soon the sound of rolling wheels, speeding

As the last man stepped out of the building and onto the ladder rung, tongues of flame ignited his coat and boots and singed his hair.

An overhead maze of ice-covered wires challenges firemen as they attempt to extinguish a stubborn fire in downtown Pittsburgh. (Carnegie Library of Pittsburgh)

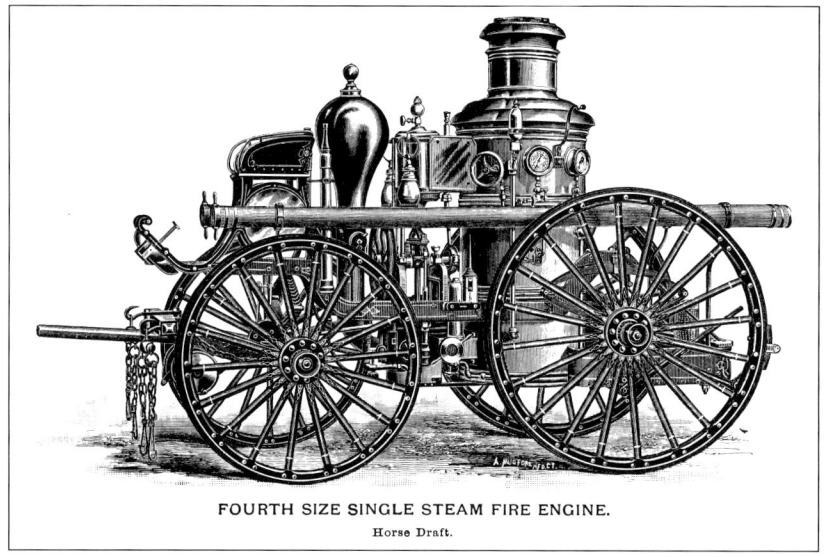
FOURTH SIZE SINGLE STEAM FIRE ENGINE.
Horse Draft.

This Fourth Size Amoskeag steam fire engine could be pulled by horse or human power. It was a popular choice to be stationed at locations that had to serve neighborhoods with narrow, winding streets. (Courtesy of the Manchester, NH, Historic Association)

FOUR-WHEEL HOSE CARRIAGE.
To be Drawn by Hand.

Manual power was the prime mover for this four-wheel hose carriage that weighed 1,300 pounds without its 1,000 feet of hose. Its use was limited to level areas and as back-up to other engine companies. (Courtesy of the Manchester, NH, Historic Association)

over the cobbled streets, was heard approaching from the north and south. The old Duquesne brigade, now officially Engine Company No. 2, slid from side to side as it tightly cornered the turn at Smithfield and Fifth, the driver off his seat, cracking the whip to urge the galloping double horse team up the steep grade to Grant Street. Close behind, Engine Company No. 3, old reliable Neptune, deftly navigated the corner from the opposite direction and fell in close behind the racing No. 2. As both pulled up in a halting slide to the smoking Court House, they dismounted, and immediately connected their hoses and proceeded to play water on the fire. No. 2 was equipped with the newer type three-inch Eureka Hose which could withstand greater pressure, thus delivering more water at greater force; the clock tower showed the time as 12:36 p.m. Crowds gathered on all four surrounding streets and jostled for position to watch the spectacle. Second and third alarms were immediately turned in. Every available piece of equipment and fireman was called out except those held for reserve. Allegheny City mustered the Columbia and Independent engine companies and, under the joint command of Chief Engineer James Crow, reported for duty along side of their Pittsburgh counterparts.

Hoses were connected to every available plug and brought to the Court House structure by passing them under gates and over walls. Once in the

This photograph shows the remains of the Allegheny County Courthouse which was destroyed by fire on Sunday, May 7, 1882. Grant Street runs parallel with the six building columns, Fifth Avenue is visible in the background at left. (Carnegie Library of Pittsburgh)

building, firemen dragged their hoses up narrow stairways and along dark corridors to reach the fire. In the center of the building was a rotunda, a circular chamber bordered by eight columns around the perimeter, which supported a vaulted roof. It was there that the center of the fire was located, already raging out of control and licking its way upward. Firemen were hampered by thick smoke and poor visibility. One firefighter described the scene as "Darker than a coal pit at midnight."

Quickly, the roof framing caught fire and spread to the dome, bringing it crashing down into the building interior. As the falling debris met the stone court house floor, a loud bell clang was heard. It was the old public bell, the first used west of the Allegheny mountain range, which had been suspended from the dome center. When the dome beams burned through and fell, so did the bell, breaking its silence of many decades.

By now, firemen inside the building realized that the situation was hopeless. Fire was spreading at an alarming rate, encircling the men with flame and falling ceilings and beams. The firefighters abandoned their positions, taking out with them what equipment and hose they could carry. In some locations, walls of fire blocked the exit paths. Firemen William Diebold, Issac Craig, William Beck, Samuel Carnahan and Michael Shenahan, members of Engine Company No. 10, found themselves trapped and had to signal their comrades from an open window for escape ladders to be raised. As the last man stepped out of the building and onto the ladder rung, tongues of flame ignited his coat and boots and singed his hair. All made it safely to the ground as the crowd cheered them on.

Others in peril were citizens, including judges and attorneys, who entered the Court House in an attempt to rescue valuable legal documents. One of these daredevils was Harry McDermott, whose grandfather was a county jury commissioner. While carrying out a box of records, he was struck on the head by a falling skylight and knocked to the floor, unconscious. Despite numerous cuts and bruises, he survived the ordeal and made a full recovery.

The same could not be said for the seat of county government; it was a total loss. What the fire left untouched, the water from many hoses used to fight the blaze flooded the building causing widespread structural damage. Sometime later, an investigation was conducted to determine the cause of the fire. The only certain fact was that no one could say with any degree of certainty exactly what had happened. Speculation abounded but centered on three possibilities. First was an errant cigar butt, carelessly discarded in a wastebasket. Other facts pointed to a lunch stand in the rotunda where a small fire was kept to heat food and beverages. Another possibility was gas lighting that was not extin-

A picture of life around the No. 2 Enginehouse shows a dual-harness hose wagon in the right-hand bay. Ceiling mounted gas lighting fixtures are ready to provide illumination for night duty. The company slide pole appears in the left-hand bay near the far wall. (Carnegie Library of Pittsburgh)

guished and set fire to the neighboring window drapery. Whatever the cause, all that remained of the Court House was a burnt-out shell filled with the ashes of its upper floors and roof. Company No. 7 received a meritorious citation for their efforts in helping to extinguish this fire.

That year also set a new record for continuous duty performed by a Pittsburgh fire company. The company was the No. 11, the old Mechanics Hose outfit, that had put in 14 hours straight at the Emory Oil Works fire on 7th Street. No sooner had the company backed into their 9th and Bingham enginehouse than the alarm sounded for assistance with a fire in progress at Sterrett's foundry. This duty kept them in service until 9:00 a.m. the next morning, a record 26 hours without a break. Six years later, No. 11 would set a record of 23 runs made in one month.

And who would have thought that another fire of major proportions would strike within 18 months? On the morning of October 3, 1883, the city awoke to a red glow in the early morning sky. The reason was a gigantic fire at the Western Pennsylvania Exposition Society grounds in Allegheny City. Flames were reflecting off the building fronts and windows to tint the sky a reddish orange color. Even the rivers were aglow, linking Allegheny City to the city with an eerie red tinge.

Burning fiercely was the Exposition Hall, a large frame and glass building containing historical exhibits constructed of flammable materials that were being systematically destroyed by the hungry flames. Both Pittsburgh and Allegheny City companies arrived almost simultaneously and began to pump steady streams of water on the building elevations; it did little good. The fire continued to advance, claiming wooden roof trusses and frame walls in its path.

When second and third alarm companies reined up, the water pressure suddenly began to drop. Hose deliveries were reduced by half. Lines were taken down to the river and maximum efforts resumed to bring the fire under control. Adjacent

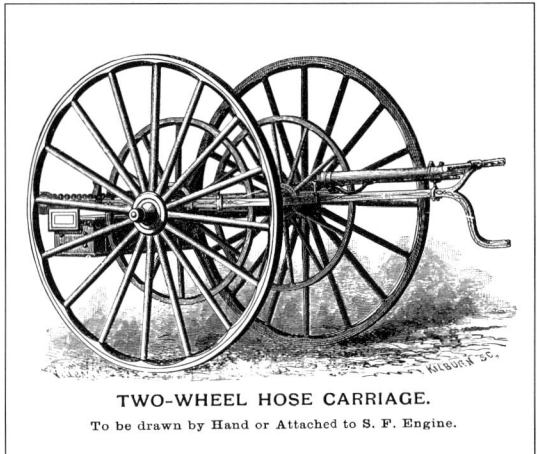

Amoskeag manufactured these two-wheeled hose carts primarily for use as trailer units attached to a larger fire engine or to support equipment. (Courtesy of the Manchester, NH, Historic Association)

TWO-WHEEL HOSE CARRIAGE.
To be drawn by Hand or Attached to S. F. Engine.

buildings were wetted down to keep flying sparks from igniting other fires. Suddenly there was a roar, accompanied by a gigantic fireball which touched off 50 barrels of oil being stored on the premises. The heat was so intense that firemen were forced to abandon the fight, pulling their hoses out after them. Fire was everywhere now, the entire complex was involved. There was no chance of extinguishing the fire. Firemen could only take up perimeter duty to keep it from spreading across adjoining streets and alleys.

When night fell on the 3rd, all that remained of the 20-acre Exposition Society grounds was a sea of ashes mixed with charred, smoldering wood. Consumed by the fire were the Machinery and Floral Halls, the Centennial Restaurant, and the beautiful Berlin Fountain which featured nighttime illumination. Stunned society members poked through mountains of debris in hopes of finding several valuable books, among them an English Bible printed in 1620 and an older German text published in 1507. They were never found. What was found, to the distress of all townsfolk, was the remains of a stable boy, John Crouse, who evidently fell asleep in an outbuilding near Exposition Hall. An inquiry into the fire fixed the cause as spontaneous combustion of the paraffine used to coat exhibition balloons. A figure of $377,747.33 was officially set as the loss to the Society buildings and their contents.

Fifteen years of fighting fires of all types and sizes gave members of the Pittsburgh Fire Department the professional edge they needed to combat the city's greatest menace. Alarm counters, indicator boards, slide poles, and automatic horse hitches were in daily use. Swinging station doors and electrical devices did their part with the ring of every alarm bell. These improvements came at a cost to the city of $525 for each station. City firemen were provided with the latest equipment, and they knew how to get the most from their machines. By now, a typical company could depart its engine house in less than a minute, under a head of steam.

Most engineers kept the fire box full of kindling or dry wood shavings, ready to light it off with an oil-soaked rag or wad of cotton waste. The station house domestic water boiler was cooking at all times. A flexible pipe connection with a spring-loaded valve gate connected the circulating hot water to and through the engine boiler heater coil. When the engine pulled out of the house, a coupling disconnected and the valve snapped shut to cut off the water flow. Enroute to the fire, precious steam was already at 10 pounds of pressure and building up a head. Once at the fire scene, a bed of channel coal was laid over the wood to make a better, hotter fire bed and keep the steam pressure up to the usual 80 or 90 pounds. To avoid boiler explosions, water levels had to be monitored very carefully, never being allowed to drop below the one-third full line on the sight glass.

Continuing to grow, Pittsburgh added three more engine companies to the department rolls. Located at Butler Street and McCandless Avenue, the No. 9 Engine Company entered service on November 1, 1885. On the 17th of the following month, newly commissioned Engine company No. 14 set up shop on Neville Street at Ellsworth Avenue in a new structure costing $24,000. Following in February 1886, Engine Company No. 13 moved into its engine house at the corner of Second Avenue and Glen Caladth Street, in Hazelwood.

In May 1886, the Fireman's Disability Fund was put into effect by the Fire Commissioner Board. This insurance plan paid a temporarily disabled fireman $30 a year. If the disability became permanent, he would receive a payment of $1,000. A like amount was paid to the heirs of a firefighter killed in the line of duty. The idea for

By now, a typical company could depart its engine house in less than a minute, under a head of steam.

Allegheny City No. 4 enginehouse was a handsome brick-and-stone structure, typical of firehouse design in the late 1800s. Curiously, most fire stations seemed to have an alarm box located across the street from them. (Carnegie Library of Pittsburgh)

this fund came about with the untimely death a year earlier of Nelson Woods, foreman of Engine Company No. 7, while fighting a fire at a Third Avenue paper and rag store. Woods was on the roof of the burning structure, trying to locate the source of the smoky fire. Without warning, the upper floors and roof collapsed, trapping him under the debris. It took his comrades four hours to pull his body from the fiery rubble.

Major structure fires continued to plague the city during the 1880s in spite of advances made in building technology. By that time, natural gas was in use as fuel throughout the city, adding another potential hazard for firemen to deal with. At the northwest corner of Wood and Fifth stood the ornate First National Bank Building, a five-story structure built around the then-modern iron frame design. The prime tenant in the building was the Western Union Telegraph Co. which had its operating facilities on the fourth floor and battery storage rooms on the fifth level. On Thursday, June 10, 1886, this handsome edifice fell victim to fire.

"Fire! Fire!" exclaimed a young boy and two men as they ran out of the building door and onto the street level. The time was 5:40 a.m. Within minutes, several engine companies were on the scene, unreeling their hoses. William Cartes, second assistant engineer; and James Stewart, third assistant engineer; pulled up to direct the attack. An explosion on the fourth floor sent a shower of window glass down on the working firemen. Flames shot out of the open window frames and thick clouds of smoke spread low over the entire area. The fire traveled up to the fifth floor and fed off the oil, which acted as insulation on the tops of several thousand glass battery jars. Within an hour, the fourth and fifth floors were a roaring cauldron of fire. Next, the red hot flames penetrated the roof construction, showering glowing debris over and onto adjoining structures.

Try as they might, the firemen were having a tough time fighting a fire that was four stories above street level. On the outside, ladders were of little help due to the tangle of overhead utility wires. The heat was so intense inside the building that you couldn't get any closer than the third floor. A major effort was directed at containment, to keep the fire from spreading to other buildings close by on two sides, separated by narrow alleys. By 8 o'clock in the morning, the fire had almost burned itself out. Saved were the bank's offices on the first and second floors. A telephone exchange located on the third floor was put out of service due to fire and water damage to equipment. The company advised patrons to use special delivery service at the post office until the equipment could be repaired. Floors four and five were totally destroyed, nothing remaining but a burnt-out shell formed by the exterior building walls. Totally exhausted, eight Pittsburgh fire companies reeled up their hoses,

Pittsburgh Bureau of Fire Amoskaeg steam fire engine in action at Water and Market Streets. Note the old postal letter box at left. (Carnegie Library of Pittsburgh)

packed up their equipment and withdrew. It was one of their finest combined efforts to date.

Investigating officials determined that the fire started in Western Union's fourth floor switchboard, the result of rats eating paraffine in the cloth insulation covering electric wiring. When bared, they came in contact with the wooden frame, touching off a concealed fire. At around half-past five in the morning, the long wooden switchboard exploded in flame, spreading rapidly to all the wooden components and furniture in the room. Despite employee efforts to extinguish it, two men and a boy were forced out by the roaring fire. Frightened, they abandoned the floor without even getting their coats and fled down the staircase to sound the alarm. In the time it took to reach the street, fire was laying waste to the entire 4th floor.

Heroic deeds of all types were performed by

Under the careful hand of its Engineer, Allegheny City's "Hope" company steam engine, a fairly new Amoskeag product, is putting in another day's work. (Carnegie Library of Pittsburgh)

The Number 1 "Hope" engine house in old Allegheny used the standard two equipment bay design flanking a center entry. Doors to the engine and hose reel equipment were bi-folding type, much the same as used in today's homes. (Carnegie Library of Pittsburgh)

city firefighters, many times in front of an assembled gathering of spectators and the curious. At Wylie Avenue and Federal Street, a four-story building containing a clothing shop on the first floor and apartments on the second floor was burning. Men of Ladder Truck "A" were helping women tenants out of the windows and carrying them down their ladders. They came to one woman who would not leave without her canary. To the applause of those watching, one member of the company brought first the woman and then the bird cage to the window, carrying both of them down to safety.

Pittsburgh's firemen continued to be tested by larger and more complicated fires. In 1887, the worst of circumstances combined to give local firefighters a most difficult challenge. Friday, August 12th was a warm mid-summer day. About 9:30 p.m. that evening an alarm was sent in from the box at the corner of Fifth and Smithfield. Nearby, smoke was billowing out from the Masonic Hall basement floor, that section occupied by upholsterer Henry Holtzman. Companies 2 and 3 arrived in an instant and began to direct water into the lower level of the four-floor brick masonry building.

In spite of the water being concentrated in Holtzman's workshop, the smoke intensified and continued to pour out of every areaway and door. Fed by a swift updraft, the flames were carried up through the building's elevator shaft. When they reached the cutting rooms of Campbell & Dick, noted fabric merchants, bulging stocks of lace, cotton, and muslins became a giant torch which ignited the entire structure. The building was now an inferno, wall to wall flame, from grade to roof, lighting up the darkened corridor along Fifth Avenue. More alarms were sent in, bringing a total of 12 engine companies to the battle. Fire Chief Samuel Evans arrived and took personal command of the struggle.

By 10:30 p.m. fire began to torch a row of dilapidated tenements behind the hall, along adjoining Virgin Alley. One company was diverted to extinguish the smoldering rowhouse roofs. It was then that the neighboring Samuel Hamilton Music Hall, superheated by the Masonic Hall flames, caught fire, its new exterior woodwork kindled by fresh coats of paint and varnish applied only

Polished and posed for the camera, Engine Company 3 displays its men and equipment for posterity. The engine is an Amoskeag, the hose reel cart also a Manchester, New Hampshire product. Company personnel include Chief John Steel standing in the center; seated from left to right are: John Herr, John Culhane, John Groetzinger, Thomas Foerst, Patrick Graham, Charles Woods, William Guntz, William McKelvey, Dan Barker, James Young and Thomas Pritchard. (Carnegie Library of Pittsburgh)

months ago. Firefighters made valiant attempts to douse the flames, using ladders to reach the high work. Time and again, they were chased back down to the street, the fire singeing their faces and necks as they descended. Still, the fire continued to grow. A gallery of more than 30,000 onlookers assembled to watch the epic, urging on their firemen, cheering as hosemen moved up to attack the shooting flames.

At 11:15 p.m. Chief Evans notified Allegheny City to dispatch whatever equipment and manpower it could spare. By midnight a total of 15 fire units surrounded the scene. The issue, however, was still very much in doubt. Another adjoining building, the eight-story Schmidt & Friday structure, was now burning, the result of firebrands carried by winds blowing up the Fifth Avenue corridor. All around, it was a mad scene, played out in the dark, illuminated by the dancing flames. There was a roar to the fire, dense clouds of smoke and scorching heat, accented by showers of red sparks, breaking glass and crashing debris amid the echos of shouting firemen, working themselves and their equipment to the limits.

Meanwhile, the Hamilton Music Hall was burning furiously from the roof down to the fourth floor. The building was regarded as a thing of beauty, with its cone-shaped roof tower. Capping the tower was a dome with a decorative lyre on the point. All of this was engulfed in flames and suddenly tilted, then spiraled downward to the street. As it fell, the flaming debris smashed in window glass on the way down, giving unneeded draft to the ever-spreading fire. City firemen continued to pour on the water from all vantage points including nearby rooftops, but it did little to slow the fire's progress.

Around 12:30 a.m., firemen confined the Schmidt & Friday Building fire to the top three floors. Early on, firefighters pulled hoses up to the Opera House roof across Fifth Avenue, and it was from this location that water was directed to contain the blaze. The fire in this building was further aggravated by broken gas pipes, which were burning away like torches. In the confusion, it took officials 30 minutes to find the correct valves to shut off the flaming gas. Firemen were very cautious when it came to fires where gas was involved. In January of the year, several men from Company No. 15 were severely burned, their first assistant chief nearly killed, when an explosion of gas was sparked while fighting a fire.

While all efforts were concentrated on the three major structures, fire was again gaining a foothold on the roofs of the Virgin Alley frame houses. Locals called this section of town 'Little Italy,' after the dozen or so Italian families who lived there. Soon, with flaming incendiaries raining down on them, they realized the coming danger and evacuated their places, taking with them what belongings they could carry. Slowly, the firemen crouched down and backed away, surrendering the old timber and tarpaper firetraps to the advancing flames. Off in the distance, the residents of 'Little Italy' gathered in building doorways, their chattels piled up close by, and watched tearfully as their homes and most of their possessions were consumed, one by one, by the fire.

On Fifth Avenue, building owner Samuel Hamilton pushed his way through the crowds and sought out Chief Evans in the maze of firemen, fire engines and yards of hose. Hamilton had previously been at the rear of his building, but as the situation became worse, he went to find the Fire Chief. Eventually he located him and discussed strategies for halting the fire's progress. As they talked, it was apparent to everyone that the building was doomed. Exterior brick walls were collapsing inward as wooden floors burned through and fell. In different parts of the building, furniture, equipment, and safes could be heard, and sometimes seen, falling to the basement as floors gave way from collapsing walls.

Allegheny City Engine Company Number 9, Springarden, showing off their Amoskeag steam fire engine and Seagrave hose cart. The dog at lower right isn't impressed, he's seen all this before. (Carnegie Library of Pittsburgh)

Allegheny City's Number 7 Silsby steam fire engine also served the City of Pittsburgh after the annexation of Allegheny in December 1907.
(Carnegie Library of Pittsburgh)

Without hesitation, fireman Harry Brobeck climbed the ice-covered ladder rungs amid thick smoke, blowing snow, and threatening flames.

It wasn't until after 2:00 a.m. on August 13th that the fires began to die down, their progress finally halted. Most of the Pittsburgh firemen had been on the line now for about five hours straight, without a break or chance to rest. A consensus among the firemen held that this fire was the worst they encountered since the riots of 1877. Perhaps that event helped in part to prepare them for this near-catastrophe. In addition to the usual fire and smoke, their work was hampered by overhead electrical wires and a tangle of fire hoses on the ground. Remarkably, although there was severe property damage, no lives were lost. Most of the credit for containing the fire in such close quarters, preventing its spread, and eventually extinguishing it was the result of dedicated, professional firework from all Pittsburgh Fire Department personnel, hoseman to chief.

Ultimately, the final count showed that both the Masonic and Hamilton's Music Hall buildings were beyond hope of repair. Both were torn down and the land cleared for new construction. The Schmidt & Friday structure, severely damaged on the upper floors by fire and on the lower floors by water, would eventually be rebuilt. Loss estimates for these three buildings alone was $750,000. All of this was the result of combustion touching off excelsior in the basement furniture shop occupied by Henry Holtzman. In a note of interest, the nearby Dispatch newspaper worked all during the fire, preparing the next day's edition. At one point the staff had to evacuate the premises as the fire crept closer; but after an hour they returned to continue their work, although ankle deep in water. On yes, the morning edition hit the streets right on time!

A blanket of snow covered the ground that winter of 1887 when an early-morning alarm came from Box 142 to No. 12 Engine Company. It was snowing hard as the boys wheeled out onto a slippery Carson Street, fighting their way through snowdrifts whipped high by howling winds. Arriving at their destination, number 1605 Carson Street, onlookers screamed that a woman was burning to death in the wood frame building. Searching the building windows, firemen saw Mary Carroll, a domestic worker, standing at one of the upper floor openings, silhouetted against a wall of flames behind her. Without hesitation, fireman Harry Brobeck climbed the ice-covered ladder rungs amid thick smoke, blowing snow, and threatening flames. Prayers were offered aloud by the crowds huddled on the sidewalks. Finally, Brobeck reached the unconscious woman and, cradling her in one arm, started back down the narrow, slippery ladder. Halfway down, the crowd gasped as the ladder began to sway. At last, he delivered her to the waiting arms of his comrades and he himself collapsed, eyes burning with smoke, face charred, and hands bleeding from cuts. It was all in a day's or night's work for a Pittsburgh fireman.

Story upon story continued to build, creating a reputation, better a legend of service provided by Pittsburgh's firemen. One of these tales involved Company No. 6 on January 15, 1888. They, along with several other companies, were heavily involved with a good-sized fire downtown at Fifth Avenue and Diamond. West End alarm box 113 sounded and No. 6 was ordered to reel-up and proceed to the scene. In a flash, the 6's were off to the west, covering the distance in less than 30 minutes, including a rail crossing stop to let a local freight train go by. On the return trip, No. 6 was notified of a fire in its district, adding an additional three miles of travel at full gallop.

The following month, on February 1, 1888, Pittsburgh's Fire Department had its name changed to the Pittsburgh Bureau of Fire. This new title placed it, along with the city police, under the newly created Department of Public Safety. These changes were the result of a late 1887 Act of Assembly, which revamped the organization of city government into three major departments. For city firemen, virtually nothing changed, except official listings, titles and stenciling of equipment. In the future, an air of politics would prevail, but even then, the basic mission would remain the same: Protect the city and its citizens from fire.

In a tribute to the department and its personnel, the president of the Board of Fire Commissioners, and well known political broker, Christopher Lyman Magee, said: "The fire department is unexcelled in its equipment and in the character and discipline of its men. The superiors of the firemen of Pittsburgh in gentlemanly deportment, in knowledge of their duties, and desire and ability to perform them faithfully, do not exist anywhere. In no other city where the same conditions exist are stubborn fires so generally confined to the places where they originate; and in no

left, **Samuel N. Evans, Superintendent of the Pittsburgh Bureau of Fire, 1878 to 1892.** (Collection of the author)

right, **William Coates was the Second Assistant Superintendent of the Pittsburgh fire bureau during the late 1800s. He would be appointed Superintendent of the bureau in 1913.** (Collection of the author)

other city is more intelligent service rendered."

High praise, but words justly deserved, for Pittsburgh firemen were always ready to answer the alarm no matter the hour or day, ever mindful that the next run might be their last. Fighting fires was a very dangerous occupation. There was always the fear of some unknown factor or circumstance rising up to cause loss of life. Firemen were especially wary answering alarms, at the fire, and returning home. Danger could lurk in the most unexpected places. The night of November 22, 1888 is a good example. Number 5 received an alarm to respond to a fire at the East End Gas Works. At the first bell the company rolled out of their beds, dressed hurriedly, and descended the pole to the enginehouse floor. Hoseman Christopher Morgan stumbled at the pole opening, made a frantic grab for the pole itself and in the process struck his head and fell lifeless to the floor below. Despite immediate medical attention, he died on the wooden floor, beside his engine.

A burning barrel factory at the Standard Oil Co. facility on December 13, 1888, added another chapter to the growing legend. Engine Company No. 9, and others, fought long and hard to save what they could of the sprawling depot, holding down the losses to $60,000, and preventing the fire from engulfing nearby neighborhoods. It was just one of the 42 alarms answered by No. 9 in its own district, not mentioning those second and third alarms from other locales, which numbered more than 100.

Number 9 was also remembered for one of the four horses that were assigned to that station. His name was 'Billy,' said to always respond when called by name, alert, intelligent and friendly. He was a large and muscular animal, calm and at rest in his stall, but eager to get under the harness. At the first alarm gong, he immediately went to the engine front and waited to be hitched up. Billy could run fast, always eager to get to the fire. When he sighted the blaze, he would head there directly, without a pull on the leathers.

Compared to previous years, 1889 was routine with regard to the number of alarms answered by the city fire service. But Bureau of Fire members stepped in to serve as another type of disaster struck Pittsburgh. On Wednesday, January 9, 1889, a violent tornado tore through the city, smashing buildings to the ground and burying their occupants inside. Wood Street was one of the districts hit hardest, the deadly winds ripping off the fronts of buildings as it passed by. One of those responding to the pleas for help was Peter Snyder, engine stoker of Company No. 1. Bravely, and without regard for personal safety, he and other members of the fire fraternity waded into the damage in search of those still alive. It was a dangerous task, as many of the remaining building walls were unsupported and leaning as if to crash down at the slightest vibration or breeze. Yet, the firemen worked on, digging out six victims from the mountains of twisted lumber, brick, and stone.

Around 6 p.m. that evening, Snyder was summoned to a collapsed building, where a lad named Joseph Goehring was trapped inside by a fallen beam, itself buried under 15 feet of wall construction that was resting on it. Along with fellow fire fighter Otto Hauch, Synder attempted the impos-

MICHAEL HANNIGAN.

Get on to Captain Mike; he is a-sliding down the pole,
Like a reg'lar acrobatic sharp, and cuts a figure droll.
Bing! bing! the gong has waked him from his slumbering serene,
And, like a streak of lightning, he gets out with the machine.

Captain Mike's the real type of what a fireman ought to be,
Clear-headed, quick, and prompt to act where others are at sea.
There's fun in his Milesian face, and a sort of devil-may-care
Expression about his flashing eye that shows he's hard to scare.

He tackles the hose carriage often when to drive he has a mind,
Through Smithfield street he tears as if Old Nicholas were behind;
The pace he takes is awful; no one else could do the like;
Which is why the people run and yell: "Hooray for Captain Mike."

But it's at the scene of action that he best gets in his work,
Where the flames are most destructive, there he labors like a Turk,

Cheering on the boys to duty, and no human pow'r can check
His phenomenal propensity to jeopardize his neck.

In the riots of '77 a heap of property he saved;
The vengeance of a howling mob for duty's sake he braved,
He's been often hurt so badly that his hopes of life were vague,
And was swiped once by the nozzle of a measly Amoskeag.

He's a handsome chap, is Captain Mike, and well he knows it, too.
The giddy girls go crazy when his manly form they view;
But he doesn't mind the silly things who round about him prance;
He sticks to duty manfully, and cusses all romance.

The chances are that Captain Mike would hold an office high;
But he happens to be a Democrat, and that's the reason why
The lightning of promotion isn't likely him to strike;
And so he's doomed to plod along as simple Captain Mike.

(257)

Many city firemen were colorful and popular individuals. Some went beyond the usual public acclaim and rated special mention as witnessed by this 1892 entry from the book "All Sorts of Pittsburghers." Engine Company No.2s Michael Hannigan shared the text with many prominent local citizens.
(Collection of the author)

sible, to extricate the injured boy from his entrapment. Snyder took a saw and, crawling around, through, and under the web of broken timber, slowly cut the beam at a location to relieve pressure on the boy. Two firemen gently carried the youngster out of danger as bystanders cheered them on. Sadly, despite the valiant effort, young Joseph died of his injuries. The joint councils of the city later voted to reward Snyder's bravery with a medal in recognition of his courageous act. Peter Snyder's work was just another example of the fire fighter brotherhoods high regard for the value of human life.

Expansion of the city fire bureau continued during the year 1889. Another engine company was placed on duty at Penn and Lang Avenue with the activation of No. 16 in January. Important fires in 1890 included one at the Maginn Cracker Co. and the L. H. Harris Drug house fire which ignited chemicals that produced fireworks rivaling the fourth of July. In the following year of 1891, city firemen were called to save Pittsburgh's Female College, the Christ Methodist Episcopal Church, and a large warehouse containing a grocery supply concern. These types of fires, requiring major effort, were, thankfully, the exception. The common fire calls for city firemen were the single-dwelling and light industrial or commercial types. False alarms, an ever-present problem, made up the balance of alarms answered by the Bureau of Fire.

Heading into the last decade of the 19th century, the city fire department had more than tripled is size since the paid service started in 1870. In February 1892, the Bureau of Fire's roster listed 17 engine companies and 4 hook-and-ladder companies. Manning these stations were 174 uniformed firemen, supervised by 4 assistant chiefs who reported to the chief, the superintendent of the Bureau of Fire. Adding in the various clerks, storekeepers, and stable workers, the staff total was 184. Bureau personnel were alerted for duty that July as the incident at Carnegie's steel works in nearby Homestead began to unfold.

On November 21, 1892, Engine Company No. 18 was officially placed in service at 121-123 First Avenue, near the corner of Chancery Lane. That very same day, sister Engine Company No. 19 established an engine house on Second Avenue below Market Street. With the activation of these units, the downtown area received the additional equipment needed to cover the increasing number of alarms originating in that district.

That same year, the department's first chemical engine was delivered by the Charles T. Holloway Co. of Baltimore. The chemical was a combination of water and bicarbonate of soda, carried in a metal tank and, when it was activated by a charge of sulfuric acid, produced a chemical reaction delivered in a liquid stream that was twice as heavy as air. The end result was ideal; oxygen was eliminated, the fire could not burn. Pittsburgh Chief Evans planned to use this equipment in outlying neighborhoods, where fires were primarily residential and water was scarce.

Bernard McKenna took office as Mayor of Pittsburgh in 1893 and with his arrival, long-time fireman Samuel Evans departed as superintendent of the Fire Bureau. Succeeding him was a McKenna appointee, Miles S. Humphreys, who promptly retitled his position to read 'Chief Engineer, Pittsburgh Bureau of Fire.' Humphreys would prove to be a capable, progressive manager and one with great longevity as well. He wasted no time in getting down to work.

"A Fire Engine Test" was the heading on a feature column in the March 14, 1893 edition of the *Pittsburgh Times* newspaper. The man behind the test was Miles Humphreys, and this was a reflection of his 'engineer' first, 'chief' second management philosophy. Reporting the event, the *Times* went on to say: "With a new Siamese

List of Firemen Injured During the Year Ending January 31st, 1896

No. of Case.	Date.		Name.	Company.	Injury.	Cause.	Days Allowed.
365 & 372	June	29	Frank Elmer	16	Luxation of knee	Injured by horse	39
366	July	4	George McClelland	15	Fracture of rib	Fall from apparatus.	42
367	"	4	Joseph Johnston	7	Burn of left leg	Burned by fire cracker	Disallowed.
368	"	20	John T. Sheppard	11	General contusions	Fall from ladder	43
369	"	26	John Baker	17	Wrench of back	Driving team	Disallowed.
370	Aug.	4	Frank McClelland	12	Incised wound of hand	Cut by falling glass	Not presented.
371	"	5	George Sylvis	11	Punctured wound of foot	Stepping on nail	4
373	"	21	Joseph Sloss	1	Sprain of shoulders	Exercise with sand bag	18
374 & 376	"	26	John Hilbert	12	Sprain of back	Lifting ladder	2
375	June	19	James Green	14	Fracture of rib	Fell from building	5
377	Feb.	17	A. R. Morris	6	Sprain and contusion of knee	Mis-step from engine	65
379	Sept.	6	George N. Day	1	Lead poisoning	Working at painting	Disallowed.
380	Oct.	3	Thomas H. Kerr	4	Burn of foot	At a fire	7
381	"	10	William H. Park	H. and L. Co. "D."	Punctured wound of foot	Stepping on a nail	8
382	"	18	Hugh Bracken	12	Sprain of hand and wrist	Being pulled from engine	7
383	"	29	Daniel Campbell	4	Punctured wound of foot	Tramped on a nailed at fire	14
384	Nov.	28	John S. Hart	23	Sprain of right ankle	Fell from step ladder	21
385	"	29	Peter Wilkinson	16	Severe contusion of testicle	Fell from engine	11
386	"	30	James H. McAleese	19	Sprain of right wrist	Fall in engine house	23
387	Dec.	3	Edward S. Hughes	2	Asphyxiated	Overcome by gas	7
388	"	15	George T. Heck	18	Contusion of knee	Collision with electric car	11
389	"	15	A. V. Burns	18	General contusions	Thrown from apparatus	17
390	"	15	Daniel Gallagher	18	Contused wound of thigh	Thrown from apparatus	3
391	Dec.	15	Thomas Sloan	18	General contusion and scalp wound	Thrown from apparatus	14
392	"	30	George A. Wright	16	Severe sprain of knee	Fall from trestle	25
393	Jan. 31, 1896		Andrew V. Burns	18	Sprain of ankle	Overturning foot	Held over.
394	Dec. 15, 1895		Thomas Fierst	18	Severe contusion and sprain of left side	Collision with electric car	20

connection, the invention of Chief Humphreys, water was thrown over an eight-story building. Those present from the Fire Bureau were the Chief and Assistant Chiefs John Steel, William Coates and James Stewart. Engine Companies 2 and 3 were used, the first taking position at the corner of Smithfield Street and the other at Grant Street. Three hundred feet of hose were used by each engine and brought midway into the block where each was connected with a four-way Siamese, 3 1/2 inches in diameter, to which a hose pipe with a 1 3/4 inch nozzle was attached. A plug pressure of 65 pounds was secured, raised to about 240 pounds by the engines, both of which carried about 95 pounds of steam. A stream was forced clear over the top of the six-story Bindley Building and the eight story Telephone Building. The four-way Siamese is an idea of Chief Humphreys and had never been tried before."

That same year, Chief Humphreys welcomed Engine Companies 20 and 21 into the Fire Bureau family. No. 20 Engine Company was located on Mt. Washington at the corner of Grandview Avenue and Sweetbriar Street, and opened its doors on July 5, 1893. Walter Avenue at Proctor Alley in the 31st Ward was the site of the No. 21 Engine Company, also put into service on the same day.

Pittsburgh's fire stations continued to grow in number with each passing year. City management recognized the safety and protection of the public and local commerce as high priorities. Yearly budgets were prepared with sufficient funding to operate and expand the city fire protection system. Chief Humphreys kept a continuous stream of requests before council for the funding of additional equipment, building new enginehouses and hiring more firemen. During the mid-1890s, the cause was aided by political continuity, always vital to passing budget requests. Henry P. Ford, who was president of the Select Council from 1890 to 1895, was elected as Mayor and served from 1895 to 1898. He believed in a strong, modern fire department and, along with Humphreys and men of the same conviction, helped to expand the Bureau of Fire service.

During the period 1894 to 1898, five new engine companies and three hook-and-ladder companies were added to the Bureau total. Engine Company No. 22 entered Bureau service on March 6, 1894. It was quartered at Fernleaf and Halibut streets in the 27th Ward. Next came Engine Company No. 23, stationed at the former enginehouse and town hall of Brushton Borough. When the city annexed the borough, this building, located on Tioga Street near Brushton Avenue, became city property; No. 23 was commissioned there on May 13, 1895.

Year 1896 brought one new engine company and a new hook-and-ladder company to the fold. On Thursday, October 1, 1896, the personnel and equipment for Engine Company No. 25 moved into the building at 3339-3341 Penn Avenue. Just one month earlier, on September 1st, Truck Company "B" joined the Bureau's force and moved into quarters with Engine Company No. 8 at Highland Avenue and Broad Street. Its equipment was a new second-class Hayes aerial ladder truck with 75-foot extension capability. It was built by the LaFrance Fire Engine Co. of Elmira, New York.

Expansion would double the following year

The life of a Fireman was one filled with the possibility of injury as evidenced by this 1890s list. Types of injuries were wide ranging as were their causes. (Collection of Ed Ross)

Engine House Number 14, Neville Street near Ellsworth Avenue. The company was placed in service during 1885. Truck Company "D" shared the same structure when it was assigned there on July 22, 1897. (Collection of the author)

Monthly Summary of Fire Service During the Year 1895.

MONTH.	Regular Alarms.	False Alarms.	Still Alarms.	Total No. of Alarms.	Miles Traveled.	Feet of Hose Laid.	Feet of Ladders Used.	Time. *H. M.
January	46	0	16	62	281½	23,650	931	206 20
February	72	0	34	106	498½	38,950	1,125	350 45
March	56	0	28	84	362½	30,650	1,657	241 35
April	41	1	31	73	299½	24,400	636	177 35
May	49	0	24	73	388¼	20,600	386	186 30
June	44	2	16	62	360	25,750	746	227 30
July	51	1	27	79	400	29,250	1,311	197 25
August	46	1	19	66	341½	25,625	876	192 20
September	32	0	24	56	226¼	11,200	202	115 55
October	77	2	37	116	558	42,850	1,225	261 20
November	47	0	21	68	328¾	11,750	348	129 55
December	48	0	26	74	320¾	17,000	535	127 55
Total	609	7	303	919	4,365¼	301,675	9,978	2,415 15

*H., Hours; M., Minutes.

October and February were the peak months for Pittsburgh Bureau of Fire work during 1895. It is remarkable that out of the 919 alarms recorded, only 7 were false alarms. (Collection of Ed Ross)

of 1897, adding two more engine companies plus more ladder trucks. On a brisk, winter January 11th, Number 26 Engine Company's roster was notified to report to their new duty station at the corner of Webster Avenue and Wandless Street. Five days later, the crew of newly activated Engine Company No. 24, with two horses, backed their new third-class, double-pump, Amoskeag engine into the garage at Ward and Wilmot streets.

Truck Company "H" waited until the spring to join up with Engine Company No. 5, uptown. It was May 14, 1897, as their 50-foot straight-frame ladder truck parked along side of No. 5's Amoskeag steam pumper. Rounding out the new quartet was Truck Company "D," which shared space with Engine Company No. 14. Their equipment was a hometown built service truck manufactured by the William H. Leonhard Company of Pittsburgh. It was delivered by its double-horse team in time for Truck Company "D" to arrive on time at its Neville Street and Ellsworth Avenue address, on July 22, 1897.

Despite all the new additions to the Fire Bureau, still more would be needed in the future. Pittsburgh was growing at a rapid rate. By 1898, the population was nearing 300,000 persons, and the city covered an area of almost 28 square miles. There were over 120 hotels and inns listed in the local city directory, with more being added. The No. 27 Engine Company was the lone newcomer to the Fire Bureau in 1898. Its new building was located at the corner of Lincoln Avenue and Renfrew Street, and was ready for occupancy in the early spring. On May 2nd, the company moved in with its third class Amoskeag pumper, now 15 years old, having started life with Engine Company No. 6.

Major city fires marked these last three years of the nineteenth century. Pittsburgh's firemen were up to the task, as usual, aided by the new men and equipment added to the department during the past five years. But, higher prices would have to be paid in order to keep fires from destroying the city. A fire at Joseph Horne's store and the T. C. Jenkins Building at Penn Avenue and Cecil Way on May 3, 1897, cost the life of one Pittsburgh firefighter. Two more city firemen lost their lives on August 26th of the same year trying to save the Edmundson & Perrine warehouse on Smithfield Street. Panic broke out in the financial community due to fires at the Pittsburgh Stock Exchange and the Union Trust Company.

There was no relief the following year as an ammonia gas explosion during a fire at the Chautauqua Lake Ice Co. also wiped out the Hoeveler Storage Co. The date was February 9, 1898. Eighteen citizens met their eternal destiny in that conflagration when they crowded in close to the fire scene and were decimated by the blast. Within 30 days, two more people would succumb

Detailed List of Fires During the Year 1895.

Date	Time A.M.	Time P.M.	Station.	Location.	Style of Build'g	Owner.	How Occupied.	Cause of Fire.	Loss.	Insurance.	Ins. Paid.
July 4	......	8.24	637	Under Lincoln av. bridge	1-story frame..	George Finley......	Dwelling..........	Fire works....			
4	......	9.04	515	3407 Charlotte street......	2-story frame..	William Zinner....	Dwelling..........	Fire works....			
4	......	9.15	525	128 Forty-third street.....	2-story brick...	George Fox.........	Dwelling & store	Fire works.....	29 87	2,000 00	29 87
4	......	9.38	43	1114 Penn avenue..........	3-story brick...	John McMahon.....	Dwelling & store	Unknown	127 00	7,000 00	127 00
5	......	3.18	92	4, 6, 7, 8 Sachem alley...	4 3-story brick	John Epley,........ Max Goldberg...... Fred Beitler.........	Dwelling and warehouse.....	Unknown			
8	9.43	......	412	Rear 2848 Carson street.	2-story frame..	John Yeager.	Cooper shop......	Burn' grubbish			
8	......	5 51	452	53 Gregory street..........	2-story frame..	Elizabeth Neimer.	Dwelling.	Chimney			
8	......	7.57	376	2005 Larkins alley.........	2-story frame..	Phillips estate....	Dwelling..........	Burning straw			
9	......	11.04	92	No. 7 Sachem alley........	3-story frame..	Max Goldberg	Warehouse.......	Ruins from other fire.....	590 43	1,000 00	590 43
11	3.02	......	114	1325 to 1331 Liberty ave.	2-story frame,.	J. G. Weir & Son.	Wagon works ...	Unknown	4,989 99	6,700 00	4,989 99
11	3.08	......	114	1325 to 1331 Liberty ave.	2-story frame..	J. G. Weir & Son.	Wagon works ...	Unknown	125 00	4,500 00	125 00
12	1.04	......	168	Cubba-You-Quit Alley..	1-story ir'ncl'd	Booth & Flinn Ltd	Brick works......	Torch explos...			
12	......	9.45	94	4 Reed street................	2-story brick....	Jacob Delbor.......	Dwelling..........	Lamp explos'n			
13	1.19	......	91	23 Bedford avenue........	3-story brick...	Peter McCoy.	Store & dwelling	Unknown			
17	8.29	......	84	606 to 612 Fifth avenue..	4 4-story brick	R. T. McGeagh...	Stores & tenem't	Unknown	6,470 00	16,000 00	6,470 00
17	8.36	......	84	606 to 612 Fifth avenue..	4 4-story brick	R. T. McGeagh...	Stores & tenem't	Unknown			
17	......	2.43	559	3504 Penn avenue.........	2-story brick...	Andrew Riemann.	Dwelling..........	Tar boil'g over	121 35	3,000 00	121 35
17	......	5 37	215	Second ave., Keystone Mill..................	1-story frame..	Jas. McCutcheon.	Shear shop......	Unknown	342 69	34,000 00	342 69
18	......	3.37	553	Rear 3317 Ridge street...	1-story frame..	S. S. McIntyre.....	Chicken coop...	Unknown			

This detailed fire list for early July 1895 shows a variety of building types, occupants and fire causes. Then, as now, fires, fireworks, and the Fourth of July seem to go hand-in-hand. (Collection of Ed Ross)

to fire in a tenement at 614 Webster Avenue. All of these fires took place in addition to everyday routine alarms where stoves overheated, or chimney sparks ignited a nearby roof.

Joining in the fight to put out Pittsburgh's fires, Engine Company No. 28 came along in the last months of the century and would be the final engine company to be activated in the 1890s. Its enginehouse was located at the corner of Filbert and Ehner streets in the 20th Ward. The structure was of typical fire house design, on a corner, where two streets intersected, which permitted each of two companies to leave independently of the other. Main floors were constructed with the equipment stall on the right as faced from the main street. On the side street, the engine bay was on the left. They connected at the rear, forming a sort of upside down letter 'L'. A kitchen, dining room, smoking room, and storage closets completed the first-floor plan. A dormitory style bedroom and bathing facilities were located on the second floor. No. 28 moved in on November 3, 1899, and shared space with Chemical Company No. 2.

Fighting large and complicated fires in tight city neighborhoods continued to be a significant part of the Pittsburgh fireman's life. There was no letup in the year 1899 with the month of January starting things off. On the 30th, a cold and snowy day, fire started in the warehouse floors of the Spear Furniture Co. at the corner of Penn Avenue and Garrison Alley. It traveled in rapid fashion to the next-door York Biscuit Company factory. Fire-

Fire Bureau Storekeepers had to maintain a diverse inventory for the department as witnessed by this 1895 List of Property. Brooms, brushes and oil were some of the more popular items. (Collection of Ed Ross)

18   *Annual Report of*

Property in charge of Storekeeper.

25 bales hay, 12 sacks oats, 26 bushels bran, 200 pounds oil meal, 6 bales straw, 12 hay forks, 3 stable forks, 2 stable shovels, 7 engine shovels, 9 rakes, 16 pounds cotton wick, 3 quires emery cloth, 23 stable brooms, 25 house brooms, 13 corn brushes, 14 horse brushes, 24 feather brushes, 10 window brushes, 4 dust brushes, 13 whitewash brushes, 24 scrub brushes, 9 dust pans, 125 pounds ivory soap, 100 pounds common soap, 200 pounds castile soap, ½ barrel tripoli, 40 pounds oxalic acid, ¼ barrel polishing oil, 1 barrel carbon oil, 1¼ barrels Eldorado oil, ½ barrel cylinder oil, 10 gallons lard oil, 15 gallons linseed oil, 5 gallons spirits of camphor, 3 gallons alcohol, 5 gallons aqua ammonia, 4 gallons roofing paint, 5 pounds vasaline, 5 pounds lard, 10 feed measures, 30 whips, 24 cow hide whips, 20 curry combs, 15 horse scrapers, 16 breast snaps, 24 pole snaps, 12 bit snaps, 16 line snaps, 60 halter snaps, 4 halter chains, 13 bits, 11 new torches, 9 old torches, 5 gongs, 6 brass oil cans, 18 reducers, 1 male screw, 40 hose straps, 26 hose gaskets, 54 dozen suction gaskets, 9 suction straps, 3 bales cotton waste, 2 bales sponges, 3 dozen chamois skins, 23 buckets axle grease, 19 stable buckets, 25 fibre buckets, 15 bushel baskets, 13 salt sacks, 350 empty sacks, 86 mops, 165 boxes extinguisher charges, 2 new extinguishers, 60 feet extinguisher hose, 300 feet steam hose, 3 gallons arnica, 33 cans hoof ointment, 8 hatchets, 12 hatchet handles, 40 axe handles, 10 pick handles, 8 sledge handles, 5 axes, 4 sprinkling cans, 56 cans, 4 oil tanks, 2 manure cans, 1 pump, 1 gallon measure, 1½-gallon measure, 1 quart measure, 3 funnels, 2 pairs scales, 1 barrel disinfectant, 1 web mosquito bar, 6 30-inch grate bars, 10 27-inch grate bars, 8 25-inch grate bars, 12 24-inch grate bars, 5 stove rings, 1 box flues, 1 box brass springs, 5 pounds wire springs, 18 2½-inch valves, 16 2¾-inch valves, 28 3½-inch valves, 100 3¾-inch valves, 60 3⅞-inch cups, 40 4-inch cups

ENGINE COMPANY No. 3.

Located on Seventh avenue, near Cherry alley.

With this company is a first-class crane neck piston engine, with 4⅝ inch double pumps; diameter of steam cylinders 7⅝ inches, with 8-inch stroke. It was built by the Manchester Locomotive Works, Manchester, N. H., and placed in service August 28, 1889. It is drawn by two horses, and weighs 8,730 pounds. It is attended by a four-wheeled hose carriage, drawn by two horses, capable of carrying 1,000 feet 2½ inch hose, and weighs 4,400 pounds. It was placed in service January 1, 1889.

ROSTER OF COMPANY.

Badge.	Name.	Position.	Former Occupation.	Residence.	Age.	When Appointed.
...	George W. King	Captain	Bricklayer	443 Second avenue	44	May 1, 1874
...	William McWhorter	Lieutenant	Puddler	2120 Penn avenue	30	October 1, 1892
30	George Brenneman	Engineer	Painter	72 Wylie avenue	40	Novem'r 1, 1889
31	Joseph Geis	Driver	Molder	330 Liberty avenue	31	May 24, 1893
32	John Green	Driver	Driver	1537 Webster avenue	40	October 1, 1879
33	George Stoudt	Stoker	Barber	29 Miltenberger street	38	August 2, 1882
34	Harry Holt	Hoseman	Laundryman	168 Howe street	31	October 7, 1891
35	James Kane	Hoseman	Puddler	2319 Forbes street	28	March 30, 1895
36	George Hughes	Hoseman	Fireman (R. R.)	469 Webster avenue	35	October 1, 1892
37	Michael Daley	Hoseman	Laborer	18 Sixth avenue	40	Dece'r 13, 1892
38	St. Clair Crawford	Hoseman	Heater	1017 Bingham street	35	Febru'y 24, 1885
39	Thomas O'Neill	Hoseman	Boiler maker	Stanton avenue	38	October 1, 1892

At age 28, Hoseman James Kane was the youngest member of Engine Company No.3 in 1895. Except for Drivers, former occupations and age appeared to have little to do with job assignment within the fire bureau. (Collection of Ed Ross)

above, **Engine Company No.2 awaits the lash-up of its suction hose before the return trip to their engine house. A recent rain has left the city's dirt streets a sea of mud.** (Collection of the author)

right, **Vests and moustaches were the order of the day for this early Pittsburgh ladder truck crew.** (Collection of the author)

men had difficulty extinguishing the smoky blaze, which was fed by cloth goods from the Spear Building and baking supplies from the biscuit plant. To make it more difficult, the fire could be fought only from the Penn Avenue elevation. The adjoining alleys were so narrow that firefighters could barely pass equipment, let alone set up hoses and ladders.

Before winter was over, the Germania Bank at the corner of Wood and Diamond erupted in fire, the result of an ashtray being emptied into a wastepaper basket. Within a very short time it seemed as if the fire had located every scrap of paper in the place. After the fire was finally put out, bank officials were horrified at the extent of the damage. What the fire didn't consume, the fireman's hose water turned into a soaking mass of wet pulp. It was never reported whether any currency was lost.

Old man fire continued to thrive during the winter season, whether from the increased use of stoves and fireplaces or due to carelessness. December 1899 featured two fires, four days apart, at institutions of learning. First to sound the alarm was the Western Pennsylvania Institution for the Instruction of the Deaf in Edgewood on the 14th. This was a doubly dangerous situation due to the inability of the student population to hear the fire bells. The alarm was communicated by the staff, using hand signs and written notices on classroom blackboards. Thankfully, everyone evacuated the premises safely.

Following on the 18th of December, St. Michaels School on the South Side caught fire. City firefighters responded in numbers and quickly went about getting the children out of the building and staging their pumper engines for the battle. Second alarms were turned in for additional help to confine the fire to the school building. By nightfall, the firemen had finished their work and headed back to their enginehouses. As they departed, cheers broke out from the crowd of parents and children assembled. The firemen were not sure whether it was for a job well done or for the fact that the Christmas school vacation was beginning early this year.

In the last year of the century, one could now look back in time over 100 years of Pittsburgh fire department history and dedicated service. This long path of progress was made possible by technology and the unbroken line of city firefighters, from the citizens with wooden buckets of the 1790s to the men and mechanized machines of the 1890s. Through the passing months and years, there was one constant in the history of Pittsburgh's firemen, and that was 'change.' As time moves forward, so do people and technology. They all change, come and go, each adding to the sum or work of those in the past. There would be new changes coming as the old calendar was about to give way to the new centennium.

Chief Humphreys had given prophetic insight to the years about to come as they would relate to firefighting. He stated publicly, in a newspaper article: "I have no doubt that there will be other forms of power that will, in the future, be used to propel our fire engines to the fire scene. These prospective improvements are, however, in the experimental stage, but have not successfully been applied to the manufacture of fire engines." Little did he realize the truth of his statement or think that he would be one of those responsible for making the new technology a working reality for the Pittsburgh Bureau of Fire. ☆

Number 2 Company's hose carriage was a four-wheel type, built by the Manchester New Hampshire Locomotive Works. This unit typically carried 1800 feet of 2 1/2 inch fire hose. (Collection of the author)

Captain George W. King of Engine Company No.13. His fire service started in 1865 as a volunteer with the Relief company. His career was marked by many heroic deeds and brushes with the grim reaper. (Collection of the author)

Riding atop of Pittsburgh Fire Department Engine No.1, this magnificent metal lamp with its red and blue leaded glass lenses let all citizens know, night or day, that the old "Eagle" brigade was still at work on the fire line. (Collection of the author)

Chapter 4

New Century, New Technology 1900-1915

With the coming of internal combustion power, horse drawn equipment rides into history.

Of all the modern technology in use at the beginning of the 20th century, none would become more important to Pittsburgh's fire bureau than the miracle of internal combustion. True, the telegraph, telephone, and electricity were indispensable tools in the course of everyday fire work, but the gasoline-powered engine would revolutionize firefighting equipment design, manufacture, and operation. By the year 1900, the idea was already 15 years old, first seeing the light of day when Carl Benz drove his three-wheeled buggy, powered by a four-cylinder gasoline fed-motor, out of a German garage in 1885.

America was not far behind, as first John Lambert of Ohio in 1891 and then the Duryea brothers of Massachusetts in 1893, built passenger-carrying vehicles that were equipped with a gasoline-fueled propulsion motor. The concept mushroomed quickly and, by the turn of the century, there were approximately 2,000 vehicles in use that utilized the petroleum fueled internal-combustion engine. Most of these were used for personal transportation, but experiments were being conducted to evaluate their potential for commercial use. In that small number of what was essentially motor trucks, the idea of using the powerplant to operate and propel fire engines was born.

Resistance to the new technology was widespread and deeply rooted. Manufacturers of steam and electric motors formed the bulk of the opposition to the new idea and did much to discredit its potential value. While the new industry tried to establish itself and promote the new design, existing old-line fire-equipment manufacturers countered with new innovations of their own. This fact in itself would prove to be of great benefit to large metropolitan city fire departments like Pittsburgh's.

Prominent fire-equipment manufacturing concerns included, among others, American Fire Engine Co., Charles T. Holloway & Co. of Baltimore, Maryland; Fire Extinguisher Manufacturing Co. of Chicago, Illinois; Thomas Manning, Jr. & Co. of Cleveland, Ohio; and the two industry giants, La France and Amoskeag, a product of the Manchester Locomotive Works. In order to meet their competition, these companies were modifying their existing equipment designs while forging ahead with plans for new and improved models. Self-propelled engines of greater pumping capacity were the universal goal throughout the industry. Preliminary prototypes featured a two-wheeled front truck steering mechanism that replaced the horse drawn-hitch assembly. The driver still had his high vantage point seat, but instead of leather reins, he steered the engine with a wheel positioned between his legs. Behind the driver, the engine frame neck attached to the rear of the truck, using a pivoted connection. Only the horses were missing, which immediately improved the driver's field of vision.

Propulsion methods ranged from steam-driven mechanisms fed by a tap off the engine boiler to electrical motors powered by lead acid storage batteries. The latter proved to be a weighty problem due to the number of batteries required to propel the heavy machine. Operating range was also limited, a factor that confined their travel to inner-city areas. While all of these matters were being worked out, horses continued to be the tractive mainstay in Pittsburgh's firehouses. Their jobs were secure for the time being, but the proverbial handwriting was on the station house walls. Horses were on their way out to pasture.

Like the mechanical equipment, horses required close and constant care. Pittsburgh veterinarian, Dr. J.C. McNeil, was kept busy tending to ailments like kidney disease, arthritis and lung problems while tending to a constant string of everyday injuries like bruised ribs, sprained legs, and cuts. The firemen themselves played a large

opposite page, **Engine Company No.19 poses on Water Street, in front of the new engine house it shared with the No.1 Company. The pumper, an extra first size American-LaFrance unit, was delivered to the department in June 1900. It was the job of these three horses to hurry its 10,735 pounds along to the fire scene, Date of this photo is circa 1904.** (Willie E. Patterson, Photographer, Collection of Robert W. Lewis)

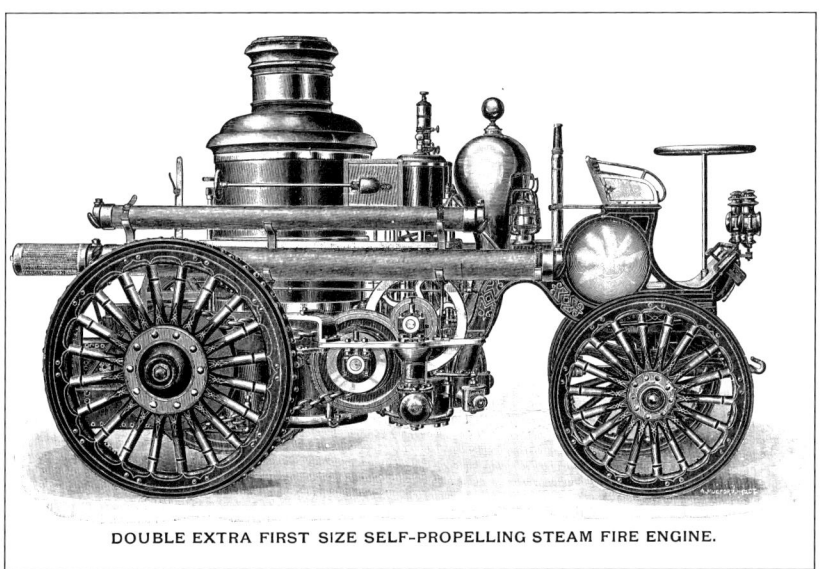

Pittsburgh's Fire Bureau received its first self-propelled steam fire engine in 1900. This giant Amoskeag unit was over 16 feet long and weighed 17,000 pounds. (Courtesy of the Manchester, NH, Historic Association)

Miles S. Humphries, Chief Engineer, Fire Chief, Pittsburgh Bureau of Fire during the years 1893 to 1913. (Collection of the author)

part in the care of their horses, particularly feeding and currying.

Firemen did not have to wait long for the fire bell to ring in the first fire of the year 1900. On New Year's Day, over a dozen different fires broke out all over the city. The following day brought no rest, either. A gas explosion and subsequent fire wiped out eight homes on the Mt. Washington hill top district. January 3rd brought with it a fire at the Wickersham School on the South Side. Fortunately, the building was vacant for the Christmas holiday recess.

Barely 18 days later, the city fire lads received a major call to duty. Shortly before midnight on Saturday, January 20th, all downtown companies were alerted for work at the multi-story Simpson Gas Appliance Co. at 26th Street and Liberty Avenue. As the first engine company drew up to battle position, gas company watchman John Simpson came running out of the burning building, his clothes afire from top to bottom. Quick thinking by the firemen smothered the flames, extinguished his burning garments and moved him out of harm's way. Due to the size of the blaze, second and then third alarms were immediately telegraphed in. At the height of the fire, more than a dozen hoses played their streams of water into the burning shell. The deluge only managed to keep the fire from spreading, a relief to worried families who lived close by. Some, afraid of the fire spreading to their homes, stood by with belongings in hand, watching the firemen work. By noon on the 22nd, the fire was out.

Nineteen hundred would continue to be a year of big fires fought along the cramped streets and narrow alleys of downtown Pittsburgh. April 7th continued the chain of alarms, as an advanced fire obliterated the Joseph Horne dry goods store. A large quantity of combustible goods contributed fuel to the burning fire, and it grew beyond the fire departments capabilities when they arrived to fight it. Five days later, three people lost their lives in the deadly fire at the Armstrong-McKelvy Lead and Oil Co. building located at the corner of Wood Street and Second Avenue.

A pair of fires in June 1900 kept firemen busier than they wanted to be. Half of the buildings in the 200 block of Fourth Avenue were threatened by a June 22nd fire, centered at the Eichbaum printing offices. More serious was the June 29th alarm answered for the Best Manufacturing Co., located at 28th Street and the Allegheny Valley Railroad. Firemen David Williams of No.15 company was killed when he couldn't outrun a falling building wall. Thirteen others were injured in the fire at this brass parts factory.

Times didn't get any easier for Pittsburgh's hosemen when the next test of human courage and strength came on July 7th. The site was located on the congested lanes of Fifth Avenue, downtown. Quickly, as if one, three establishments began to belch smoke and fire. Some think the fire started in the lofts of the Oliver McClintock Carpet Co. It mattered little, for in a flash both the Goodard, Hill & Co. jewelry firm and a china and glass store operated by T.G. Evans were heavily involved by the flames. When it was finally extinguished, the property loss, while great, paled before the personal toll of six firemen killed. No. 4 company lost Daniel Campbell, Stewart Burns, Clare Crawford and John Griffen. No. 11's fatalities were Max Butterbach and John Lewis. All six perished together when the floor they were working on collapsed, and sent them plunging down

above, **Number 32s Engineer makes a valve adjustment as bystanders, including the dog, policemen, and the nattily dressed gent at the left, stand around.** (Collection of Richard L. Linder)

left, **Firemen battle a smoky fire from atop a Hayes Ladder and Combination truck. Support wagons are assisting their efforts. The date is October 1, 1900.** (Carnegie Library of Pittsburgh)

through several floors to the basement.

Although the town firemen were, as a general rule, level-headed and mature, certain things could startle them or upset their concentration to the point of distraction. Having a black cat cross your engine's path or walking under ladders at fires were sure to cause anxiety, but major fires that occurred on the anniversary date of earlier fires past were especially haunting.

It was on the 9th of February, 1901, when the fire telegraph notified downtown engine companies that a multiple structure fire was under way at 24th Street and the Allegheny Valley Railroad tracks. Two neighboring businesses, the Armstrong Cork Co. warehouse and the Totten & Hogg Iron Foundry were joined by a wall of flames. Continuing, the alarm board tapped out the address. It was then that someone remembered there was a fire about four blocks from that location eight months earlier that claimed the life of David Williams. Another fireman spoke up with even greater recall: this day was the third anniversary of the ice company chemical explosion that killed 18 souls.

In March of 1901, the city firemen lost one of their greatest patrons and staunch defenders as death claimed Christopher Lyman Magee. Often called the consummate politician, dealmaker or a dozen other irreverent titles by his detractors, he was, above all else, a man who loved the firemen, and they certainly loved him in return. Magee was the last chairman of the Board of Fire Commissioners and helped write the language that merged the paid fire department into the city bureau. In tribute, the 10 pall bearers who carried him to his last resting place were all Pittsburgh firemen, led by Assistant Fire Chief John Steele. Years later, when Magee's wife Eleanor died, firemen also stepped forward to serve as her pall bearers.

For sheer size, the Exposition site fire of March 16, 1901, ranked among the city's largest. Compared again most other Pittsburgh fires, it was a blaze of gigantic proportions and covered a wide area. Starting in the Western Pennsylvania Exposition Building, the fire spread rapidly to adjacent exhibit structures even before the first fire hose was laid down. Even with a second and third alarm, firemen had great difficulty confining the fire due to a sharp breeze. Not only were the exposition grounds burned out, but also, 20 homes were destroyed before the fire was declared out. One Pittsburgh fireman lost his life during this fire, the result of a wall collapse.

Fire seemed to haunt the South Side during the remaining months of 1901. A department store belonging to the George Lorch Co. on Carson Street caught fire on April 29th. Fed by fabrics and furniture, the flames ignited nearby buildings, complicating work for firemen at the scene. Several months later, an industrial fire at the Dillworth, Porter & Co., located on Bingham Street between South Fourth and South Sixth streets, spread out to envelope a four-acre site.

One location that always produced large and complicated fires was the Penn Avenue district between Sixth and Tenth Streets. There was at least one major alarm there each year. On Thursday, April 3, 1902, the record was kept intact with

One location that always produced large and complicated fires was the Penn Avenue district between Sixth and Tenth Streets.

How about a few hoses! The 1900 fire at DeNoon's Wholesale Paint and Glass Warehouse at Seventh and Liberty is almost out. Firemen and spectators alike turn and direct their attention to the cameraman. (Carnegie Library of Pittsburgh)

a call from Box 32 at 5 o'clock in the afternoon for assistance at the Baker-Williams Furniture Co. at 817 Penn Avenue. A fire in this type of establishment was always hard to fight. Burnable materials were stacked together tightly with very little aisle space between. Fires tore through the upholstery and bedding, not to mention all the wooden furniture and floor coverings. Repair and refinishing shops, with their stains, glues, and varnishes, provided accelerants to stoke the flames to superheated temperatures, carrying away everything in its path. Fortunately, the water tower brigade mobilized early and, fed by three engines, poured thousands of gallons of water into the upper stories of the building.

For Baker-Williams, it was the second fire within a year, the April 1901 event taking the life of city fireman Edward Hagmaier. The 1902 edition had a near-fatal accident that occurred some blocks away from the scene of the fire. The horse pulling Public Safety Director A.H. Leslie and his driver, Charles Seers, stumbled and fell at Penn Avenue and 16th Street. Both men were catapulted to the street, suffering cuts and bruises, but were otherwise all right.

Pittsburgh's Bureau of Fire had grown significantly in these early years of the 1900s. Both in physical size and service extent, the city fire department was growing larger. A look at Bureau statistics for the year ending 1906 told the impressive story. A total of 1,200 fires were fought by city firemen, an average of better than three per day. Firemen traveled almost 5,400 miles to answer the fire duty call. There were 35 engine companies in the department utilizing almost 100,000 feet of hose. In support, 2 chemical companies and 11 ladder companies were strategically located throughout the city. Manpower totaled 468: one chief, 6 assistant chiefs, 38 captains, 44 lieutenants, 33 enginemen, 84 drivers, 256 hose and ladder men and 6 fuel wagon men. To meet the responsibilities of maintaining firefighting service in a growing city, the fire bureau would continue to expand in numbers of personnel and types of equipment.

The annexation of Allegheny City to Pittsburgh became effective on December 6, 1907. This added 8.07 square miles to the municipal boundaries, making a combined total of 38.30 square miles protected by the Pittsburgh Bureau of Fire. Official consolidation of fire service became effective four days later, on December 10th. It was at this time that the term Greater City, or more popularly, Greater Pittsburgh was coined. Included in the metropolitan area was the downtown peninsula, east to, and including, Brushton Borough;, Allegheny City;, South Side; and tracts of land west that captured Esplen and Sheridan Boroughs.

When Allegheny's fire department merged with that of the city, its ranks were made up of 10 engine companies, 6 hose brigades and 5 hook-and-ladder companies. One chief and 3 assistant chiefs had overall command of the department with immediate supervision being vested in 18 captains. There were no lieutenants in the Allegheny City Fire Department. The balance of personnel included 10 engineers, 32 hose men, 23 ladder men, 2 fuel wagon drivers, 1 clerk, and 1 repair shop superintendent making an overall grand total of 91 men.

Allegheny was by far the largest addition and became the North Side in official records of the fire bureau. Events barely waited until the merger ink was dry to challenge the newly combined department with a devastating fire. The date was February 25, 1908, the location Boyd's Trunk Factory on Isabella Street. Strong winds from the Allegheny River fanned the multi-story fire into a raging inferno. Wind forces were so strong that they even deflected the hose streams from their

Robert McKinley, assistant fire chief, Pittsburgh Bureau of Fire.
(Collection of the author)

William Bennett, assistant fire chief, Pittsburgh Bureau of Fire.
(Collection of the author)

intended targets. One of the tallest structures on the North Side, the building was quickly overpowered by the flames, which began to spread to homes on the down side of the wind. When the firemen arrived, factory management was trying to evacuate the burning building of its workers. Those on the lower floors climbed out of windows and jumped to the ground. Workers on the upper floors had no open escape routes. It was only after the fire was out that the full extent of the tragedy was known, for not only was the building a total loss, but the bodies of 11 young girl employees also were found among the charred building debris.

By the year 1908, the amalgamated Pittsburgh Bureau of Fire consisted of the following component companies and equipment:

Engine Company No. 1

Formerly the volunteer company known as "Eagle," it was first located on Fourth Street near Liberty Avenue for a year and then moved to First Street and Chancery Lane. The company became part of the paid city fire department when it was organized on June 1, 1870. Moved to temporary quarters at the corner of Strawberry and Cherry Ways on March 19, 1906, their old property purchased for use by the Wabash-Pittsburgh Terminal Railway. On May 23, 1906, the company moved to its present location in a new engine house at the corner of First Avenue and Short Street.

Equipment consisted of one second size Amoskeag engine placed in service on December 26, 1900, and one Amoskeag four-wheel hose carriage rebuilt by the James G. Weir & Sons

A large crowd gathers to watch the April 24, 1901 fire at the Barker-Williams Furniture Warehouse. Firemen are attacking the upper level fire from rooftops across the street. At right, a water column is in position and playing its water stream into the lower floors.
(Carnegie Library of Pittsburgh)

Engine Company Number 1 tests its equipment in the afternoon sunlight along Water Street. The structure to the right is the Number 19 Engine House. (Carnegie Library of Pittsburgh)

Wagon Works and placed back in service on February 24, 1892. This carriage was first equipped with rubber tires on January 18, 1899, and then renewed with its second set of rubber tires on September 20, 1904; both were drawn by two horses.

Engine Company No. 2

Formerly the volunteer company known as "Duquesne," it was first located on Smithfield Street near Second Avenue. The company became part of the paid city fire department when it was organized on June 1, 1870. On April 10, 1900, the company moved to temporary quarters at Cherry Alley near Seventh Avenue to allow its new enginehouse to be constructed. The company moved back into the new enginehouse at the same location on December 24, 1900.

No. 2's equipment consisted of one first size Amoskeag engine placed in service on September 14, 1896, and one combination wagon with turret built by the Fire Extinguisher Manufacturing Co. of Chicago, Illinois and placed in service on June 21, 1900, both were drawn by two horses.

Engine Company No. 3

Was located on Seventh Avenue near Cherry Alley and was the former "Neptune," volunteer company. No. 3 was placed in service on June 1, 1870. The original enginehouse was torn down and reconstructed in 1894.

This company was equipped with a first size Amoskeag brand engine, manufacturer's serial number 750, which was placed in company service on May 5, 1900. No. 3's hose carriage was built by the Manchester Locomotive Works of Manchester, New Hampshire, and placed in service on January 1, 1888. It was rebuilt on August 1, 1898, by the William H. Leonhard Wagon Works of Pittsburgh; it was drawn by two horses.

Located in the same quarters with Company No. 3 was Ladder Truck Company "E". This company was put in service on June 1, 1870. Its equipment was an 85-foot Hayes Aerial Truck built by the LaFrance Fire Engine Co. of Elmira, New York and entered service on January 18, 1897. The truck was rebuilt by the James G. Weir & Son Wagon Works on October 25, 1904; it was drawn by three horses.

This broadside of Number 2's Combination Wagon shows the detail of all the equipment carried on board. The boys look at ease and dapper in their straw boaters. Even the company mascot has taken his place in this scene. (Carnegie Library of Pittsburgh)

Engine Company No. 4

Was located at 1726 Fifth Avenue between Miltenberger and Van Braam Streets and, despite some conflicting information linking it with the Neptune Company, it was formerly known as the "Relief" volunteer company. No. 4 became part of the paid fire department on June 13, 1870.

Its engine was a second size LaFrance with double pumps that produced 750 gallons per minute. Weighing in at 8,470 pounds, it was placed in service on December 12, 1894, and was pulled by three horses. Attending this company was a four-wheel Amoskeag hose carriage drawn by two horses. Weighing 4,200 pounds, it was placed in service in June 1883.

Engine Company No. 5

Was placed in service on April 8, 1872, at 2155 Center Avenue where it remained until December 1, 1903. At that time they moved into their brand new building at Center Avenue and Devilliers Street.

Engine Company 5's apparatus was a two-horse third size Amoskeag double pump unit with a capacity of 550 gallons per minute. Weighing 6,400 pounds, the engine was delivered August 24, 1884. The hose carriage unit was a four-wheel Amoskeag drawn by two horses. It weighed 4,400 pounds and was placed in service on that same date.

Truck Company "H" was located in the same house with Engine Co. 5 owned a straight frame city truck with a 50-foot ground extension ladder. It was drawn by two horses and was placed in service on May 14, 1887.

A state of the art, modern fire fighting brigade, vintage 1915. Here is the Number 4 Company set up in front of its engine house at 1726 Fifth Avenue, uptown. Both the engine and truck bear the Maltese cross American-LaFrance emblem. The lunchroom spectators at the right interrupt their meal to join in the photo. (Carnegie Library of Pittsburgh)

Engine Company No. 6

Located at 44th and Calvin streets, it was placed in service on May 30, 1873. Previous to that date, it was called Hose Company No. 10, a descendant of the old Lawrenceville Hose Company.

With this company was a third size Amoskeag double-pump engine, serial number 617, built by the Manchester Locomotive Works of Manchester, New Hampshire, in May 1886. This engine originally went into service at No. 15 company in

Pride of Engine Company No.6 was this 1913 Knox chemical and hose car. It took a hefty turn on the engine crank to turn the four cylinder motor over for starting. No doubt those treadless tires had a short duty life, the front one already showing a long gash from driving on Pittsburgh's rugged streets. (Willie E. Patterson, Photographer, Collection of Robert W. Lewis)

May 1886. On January 10, 1890, it was transferred to No .6 company and rebuilt with a new boiler in July 1899 at the Manchester Works. It was placed back in service for No. 6 in October 1899. The apparatus weighed 7,200 pounds and was drawn by two horses.

A four-wheel Amoskeag hose carriage was placed in service with this company in August 1883. William H. Leonhard rebuilt the reel in May 1894; the carriage weighed 4,500 pounds and carried 1000 feet of 2 1/2" hose.

Engine Company No. 7

This company went into service in June 1870 and was the original "Independence" volunteer company. Its original enginehouse was torn down in 1905, and they moved to occupy quarters at 23rd Street and Penn Avenue.

Equipment with No. 7 included a second size Amoskeag engine, manufacturer's serial number 452. It was placed in service on June 12, 1873, at No. 2 engine company, transferred to No. 15 engine company and then to No. 7.

A four-wheel hose carriage of Amoskeag manufacture accompanied No. 7 to all fires. It was placed in service in 1879 and was drawn by two horses. On August 1, 1898, the rig was rebuilt by the William H. Leonhard Co.

Engine Company No. 8

When the paid fire department was organized, this company was assigned duty on Frankstown Avenue near Penn Avenue. In August 1873, this company went into service at Highland Avenue and Broad Street. The enginehouse was modernized and rebuilt in 1896.

A second size LaFrance engine, serial number 313, was assigned to No. 8 on December 12, 1894; it weighed 8,400 pounds and was pulled by two horses.

The hose carriage was an Amoskeag type with four wheels, and it was placed in service in 1885. The equipment weighed 4,200 pounds, including 1,000 feet of 2 1/2 inch hose, and was drawn by two horses.

Truck Company "B" was also resident in the same structure with No. 8 and used a second class Hayes Aerial Truck equipped with a 75 foot ladder built by the LaFrance Fire Engine Co. of Elmira, New York. Two horses pulled the 6,000-pound unit. It was placed in service on September 1, 1896.

Engine Company No. 9

Went into service on November 1, 1885, at Butler Street and McCandless Avenue; until the year 1875 it was only a hose company.

The equipment at No. 9 was a second-size crane-neck double-pump Amoskeag engine built in December 1867 and known as the "Vigilant." The engine was rebuilt at the Manchester Works in June 1875, and was delivered to Engine Company No. 9 for commissioning on November 5, 1887.

Company hose carriage was an Amoskeag four wheel type placed in service on November 1, 1885. The carriage weighed 3,200 pounds and carried 1,000 feet of 2 1/2 inch hose, two horses pulled this carriage.

It is the year 1913, and the West End Hose Company No.10 and crew prepares to take their circa 1907 Seagrave hose reel for an early spring run. (Willie E. Patterson, Photographer, Collection of Robert W. Lewis)

Engine Company No. 10

Located at Steuben Street and Mill Street, this engine company went into service on June 15, 1875.

With this company was a second-size Amoskeag engine with a capacity of 650 gallons per minute. This engine entered service on June 13, 1873. A boiler overhaul project was completed in 1900; this equipment was drawn by two horses.

Supporting this engine was a four-wheel Amoskaeg hose carriage that was placed in service on February 9, 1893. It was the best hose cart in the department; the unit was drawn by two horses.

Engine Company No. 11

Bingham and Ninth streets was the location for this engine company, which went into service on February 8, 1873. In volunteer days this company was known as the "Mechanics Hose Company," and was located on South 14th Street. No. 11 moved into its present quarters on August 8, 1874.

Equipment assigned to this company included a second-size Amoskeag engine with a crane-neck and double pumps. It was originally placed in service in December 1867. In 1875, the engine was returned to the Manchester Works and rebuilt. A new boiler was fitted to the engine in 1902 by the LaFrance Fire Engine Co. in Elmira, New York.

No. 11's hose carriage was a four-wheel affair, Amoskeag brand, placed in service in 1886; it was drawn by two horses.

Engine Company No. 12

Was formerly the "Walton Hose Company" in the days of the volunteers. The company was then located on Sarah Street between South 20th and South 21st streets in an old two-story frame building. Company 12 was placed in service on February 15, 1873. On January 30, 1884, the company moved into its present location on Carson Street between South 20th and South 21st streets.

The first engine entrusted to the Walton Company was the old "William Lyons No. 2," a second-size Amoskeag crane-neck type, builder number 452. The next engine assigned was a third-size Amoskeag crane-neck type with double pumps; it was pulled by two horses and placed in service in May 1884.

Assisting this company was a Champion combination chemical engine and hose wagon manufactured by the Fire Extinguisher Co. of Chicago, Illinois. Its tank had a 60-gallon capacity. Two horses drew this wagon, which entered service on November 3, 1889.

Ladder Truck Company "C" resided in the same house with No. 12 and was activated in 1873 using an old H.I. Gourley brand truck. It was replaced by a new 75-foot aerial Babcock-type truck built by the Fire Extinguisher Manufacturing Co. of Chicago on November 22, 1890. This same truck was rebuilt by the LaFrance Fire Engine Co. into a Hayes aerial ladder truck in June 1895.

Engine Company No. 13

First went into service as a hose company on November 12, 1885, and then converted to an engine company on August 11, 1886, and was quartered at Second Avenue and Glen Caladth Street.

No. 13's engine was a fourth-size single-pump Amoskeag type, producing a capacity of 350 gallons per minute, it entered service in August 1873. The engine was rebuilt from a goose-neck to a crane neck configuration in 1899. A team of two horses drew this unit.

With a gallery of spectators standing around, Engine No. 11 goes about the mundane business of pumping out a flooded building basement. (Collection of Richard L. Linder)

The Number 12 engine house surely must have won a prize for the "Best decorated engine house." Located in the 2100 block of Carson Street, South Side, this 1907 photo shows the entire company lined up in front of their handiwork. Note the figure on the roof top. (Willie E. Patterson, Photographer, Collection of Robert W. Lewis)

This gold medal was presented to Captain George W. King of Engine Company No.13 by his comrades in 1901. The reverse side reads "Presented- Geo. W. King by the members of Eng. Co.13" (Collection of Richard L. Linder)

The hose carriage, a Pittsburgh product built by the W.H. Leonhard Wagon Co., was a four-wheel variety and was pulled by two horses.

Truck Company "G" shared the house with No. 13. Its equipment consisted of a straight frame city service truck drawn by two horses, and started service on November 10, 1904.

Engine Company No. 14

On December 17, 1885, Engine Company No. 14 entered service. Its engine house was located on Neville Street near Ellsworth Avenue.

Their third-size crane-neck Amoskeag engine with double pumps was ready for duty on November 15, 1907. The 7,950-pound engine was drawn by a team of two horses.

Assisting with hose was a four-wheel Amoskeag-type carriage pulled by two horses. Its 4,500-pound weight included 1,100 feet of 2 1/2 inch hose. It was the first piece of equipment in the engine house door, arriving for duty on December 17, 1885.

Truck Company "D," with its William G. Leonard city service truck equipment, joined up with No. 14 on July 22, 1897; two horses were needed to pull the truck.

Engine Company 14 is getting ready to leave their station, probably for some official function. The engine is a Third Size Amoskeag type, all spit and polish for the occasion. The building in the background is the old Duquesne Gardens. (Carnegie Library of Pittsburgh)

Engine Company No. 15

Entered the paid fire department in 1870 as the "Niagara Steam Fire Company." Its name was immediately changed to Engine Company No. 2. In 1873, its old Amoskeag engine "Niagara" was sent to Engine Company No. 6. Identities changed again as No. 2 now became Hose Company No. 1. The company operated all over the city, answering alarms with four or five men, except for the South Side and East End. In January 1886, the company received a new engine from the department shop and was designated Engine Company No. 15.

The next engine assigned was Amoskeag No. 698, originally placed in service at Engine Company No.2 on October 3, 1893, coming finally to No. 15 on September 14, 1896. Its enginehouse was located at Penn Avenue near 14th Street.

A four-wheel Amoskeag hose carriage pulled by two horses serves this company, it entered on duty August 18, 1885.

Engine Company No. 16

Served the 21st ward from its enginehouse at the corner of Penn Avenue and Lang Avenue. Doors opened for business on January 1, 1889.

Engine equipment was a second-size crane-neck piston-type, manufactured by the Thomas Manning Co. of Cleveland, Ohio. Manufacturer's serial number is No. 114. It entered service on January 8, 1897. In August 1900, the LaFrance Fire Engine Co. rebuilt the equipment.

Truck Company "I" shared space with No. 16 and was a third-class Babcock type hook-and-ladder truck built by the Fire Extinguisher Manufacturing Co. of Chicago. It entered service on September 30, 1903, and was pulled by two horses. The following ladders were carried on the truck: One 50-foot extension type, one 40-foot extension type, one 30-foot ladder, one 20-foot ladder, one 16-foot ladder and one 10-foot ladder. It was also equipped with one turret deck nozzle. This truck was one of the last produced under the Fire Extinguishing Manufacturing Co. nameplate. With eight other old-line companies, it merged to form the International Fire Engine Co., which, in a few short years, would change its name to the American-LaFrance Fire Engine Company.

Engine Company No. 17

Entered duty as Hose Company No. 2 on January 1, 1877, at Bailey Avenue and the Castle Shannon Incline Plane. On February 16, 1891, it was assigned engine company duty at Virginia Avenue and Shiloh Street on Mt. Washington.

A third-size Amoskeag single-pump engine was the mainstay of No .17's equipment. It was built in February 1872 under builder's number 403 as the "William Phillips," and delivered to Engine Co. 5. In 1886, it was reassigned to Com-

B.J. Cawley, assistant fire chief, Pittsburgh Bureau of Fire. (Collection of the author)

top right, **A team of well cared for horses awaits the command to move out with Engine Company 14's Hose Reel.** (Carnegie Library of Pittsburgh)

bottom right, **This winter scene shows Engine Company 16 and its patriotic crew at work. The older Amoskeag engine is being drawn by a front wheel drive gasoline powered Christie tractor.** (Carnegie Library of Pittsburgh)

pany 17 and saw duty until 1898, when it was rebuilt from a harp neck to a crane-neck type; two horses pulled this engine.

Providing hose for this company was a light-weight Amoskeag hose carriage with four wheels. Drawn by two horses, it was put into service in 1873.

Engine Company No. 18

A downtown company, it was first placed in service on November 21, 1892, at number 121-123 First Avenue at the corner with Chancery Way. On April 5, 1902, it moved to its present home at Maddoc Way and Penn Avenue. Some old-timers knew this company as the "Ben Franklin."

The engine assigned to No.18 was an Amoskeag first-size crane-neck type, drawn by two horses. It entered service on March 25, 1899.

Hose service for the company was provided by a four-wheel Amoskeag hose carriage, which started life at Company No.3 in 1873 and transferred to No. 18 on November 21, 1892. It was rebuilt by T.A. Dutenberg of the South Side in 1898; two horses were used to transport this reel.

Engine Company No. 19

First located on Second Avenue below Market Street, it was placed in commission on November 21, 1892. The company was moved to a new house on 8th Street between Penn Avenue and Duquesne Way on January 31, 1902, because the lease on the old property expired. On June 13, 1906, the company moved into its new quarters at Water and Short streets. Cost of the new building was $52,000.00.

With this company was an extra first-size American type fire engine manufactured by the LaFrance Fire Engine Co. of Elmira, New York. In service date was June 9, 1900. The engine was equipped with double 5 3/4 inch diameter pumps with steam cylinders of 9 1/2 inch diameter and 9 inch stroke. Capacity was a mammoth 1,000 gallons per minute. Weighing in at a hefty 10,735 pounds, it took three sturdy steeds were required to pull it.

above, **James Connolly, assistant fire chief, Pittsburgh Bureau of Fire.** (Collection of the author)

top left, **A wealth of detail is evident in this 1901 photo of Engine Company 19's Deluge Wagon, the only one of its type ever built. Foremost is the three collar hitch arrangement. Also to be noted are, at left, the slide pole, wall telephone replete with hanging directory and decorative metal panel ceiling.** (Carnegie Library of Pittsburgh)

middle left, **Engine Company 19's crew pauses briefly for the camera while their LaFrance engine continues its work.** (Carnegie Library of Pittsburgh)

bottom left, **Engine Number 19 and crew works in the icy cold of December 12, 1903, to help extinguish a fire at the Haugh and Keenan Warehouse.** (Carnegie Library of Pittsburgh)

above, **Peter Snyder, assistant fire chief, Pittsburgh Bureau of Fire.** (Collection of the author)

Supporting this engine was one of the most unusual pieces of fire equipment ever built and it was, in fact, unique. It was called a deluge wagon and was built by the Fire Extinguisher Manufacturing Co. of Chicago. No. 19 put the rig in service on June 9, 1900. It had the distinction of being the only one of its type ever built. The design included a 24-foot water tower, 2 deck turret nozzles, each with connections for 3 lines of 2 1/2 inch or 3 inch hose. A 100 gallon chemical tank with a reel of chemical hose was constructed on the steel truck bed which was divided in the center to carry 2 lines of both 2 1/2 inch and 3 inch chemical hose. Also aboard were 2 ladders and 2 Babcocks. All told, the unit weighed in at 12,000 pounds, but could be pulled by three horses due to roller bearing wheels and rubber tires.

In an adjoining space was Truck Company "A," another engineering marvel of the era. The truck was a patented quick-raising aerial ladder type built by the American-LaFrance Fire Engine Co. Mounted on the steel channel frame was an 85-foot aerial ladder, which pivoted and extended with powered assists rather than the hand-crank methods. Also equipped with roller bearing axles, three horses could handle the duty of delivering the unit to the fire scene. It was placed in service on October 9, 1906.

Engine Company No. 20

This Duquesne Heights company was located at Grandview Avenue and Sweetbriar Street. July 5, 1893, was the date it first saw service.

Equipment for this company consisted of a Holloway Chemical Engine with dual 60-gallon tanks. The two-horse unit was manufactured by the Charles T. Holloway Co. of Baltimore, Maryland. It was placed into service on September 24, 1893.

Hose service was provided by a four-wheel carriage pulled by two horses. The 4,835 pound reel first saw service in 1893.

Engine Company No. 21

On July 5, 1893, Company No. 21 was placed in service at Walter Avenue and Proctor Way in the 31st Ward.

With this company was a fourth-size Amoskeag single-pump engine that produced 350 gallons per minute. Originally a goose-neck engine, it was rebuilt into a crane-neck type. Service for this unit dated from 1874; a two-horse team was required to pull it.

An Amoskeag two-horse lightweight hose carriage was put into service at this company on July 5, 1893.

Engine Company No. 22

This company was located at Fernleaf and Halibut streets in the 27th Ward, and service was initiated on March 6, 1894.

Assigned to Company No. 22 was a four-wheel hose wagon built by James G. Weir & Co. of Pittsburgh. Two horses were used for pulling the wagon, and it was placed in service on February 9, 1893.

Engine Company No. 23

No. 23 was set up in a building that was formerly the old Brushton Borough Town Hall and Engine House. It was built in the autumn of 1892. When Brushton Borough was annexed to the city, the building was designated as the enginehouse for Company No. 23. They moved into these quarters on May 13, 1895.

Equipment for the company consisted of a Holloway Combination Chemical Engine and Hose Wagon built by the Charles T. Holloway Co. of Baltimore. Two 35-gallon chemical tanks were installed under the rear wheels and were connected to two chemical hose reels, one carrying 900 feet of 2 1/2 inch chemical hose and the other carrying 250 feet of one inch chemical hose. Total weight of the wagon was 6,500 pounds, which was managed by two horses.

Engine Company No. 24

Resided in the enginehouse at Ward and Wilmot Street, and was placed in service on January 16, 1897.

No. 24's engine was a third-size double-pump of Amoskeag manufacture. Water output capacity was 550 gallons per minute, two horses were used to pull this unit.

Hose carriage for this company was a local product of the James G. Weir & Sons Wagon Works, whose factory was located on Liberty Avenue between Thirteenth and Fourteenth streets. It was delivered in 1898 and pulled by two horses.

This enginehouse was also the headquarters of Assistant Fire Chief Peter Snyder.

Tuesday, July 7, 1908 looks like a relaxing day at Engine Company 20's engine house on Grandview Avenue. Manhole castings on the nearby wagon indicate that some sewer and road work was in progress. (Carnegie Library of Pittsburgh)

Engine Company No. 25

Located at 3339-3341 Penn Avenue, was commissioned on October 1, 1896.

Its engine was a second-size LaFrance double-pumper type, builder's serial number 425. It was placed in service on November 8, 1899. Weight of this engine was 9,930 pounds and required three horses to pull it.

Attending this company was a two-horse, four-wheel chemical engine and hose wagon built by the Fire Extinguisher Manufacturing Co. of Chicago, Illinois. Equipment included one 60-gallon chemical tank, one 3-way siamese deck turret nozzle, 1,000 feet of 2 1/2 inch chemical hose, and 250 feet of 1 inch chemical hose. The wagon weighed in at 6,500 pounds and was delivered and placed in service on June 18, 1900.

Truck Company "F" resided in the same building with No. 25 and utilized an 85 foot Babcock-design aerial ladder truck built by the Fire Extinguisher Manufacturing Co. Apparatus included a three-way Siamese deck turret nozzle. This unit weighed 8,500 pounds and required three horses to pull it. In-service date for this truck was October 10, 1902.

Engine Company No. 26

On January 11, 1897, Engine Company No. 26 was put in service at the corner of Webster Avenue and Wandless Street.

Assigned to this company was a fifth size crane neck double pump engine built by the American Fire Engine Co. of Seneca Falls, New York. The engine was put into service on May 15, 1898.

A James G. Weir & Sons four-wheel hose carriage complimented this company's engine and came to No. 26 on July 27, 1897. Drawn by two horses, it carried 1,000 feet of 2 1/2 inch hose and weighs 3750 pounds.

Engine Company No. 27

This brigade was located at the corner of Lincoln Avenue and Renfrew Street. In service date was May 2, 1898.

A third-size Amoskeag double-pump engine, builders serial number 582, resided in this house. It weighed 6,400 pounds and was pulled by two horses. The engine arrived for duty at No. 27 in June 1883.

William H. Leonhard Co. of 218 43rd Street, Pittsburgh built this hose carriage in 1898. It carried 1,000 feet of 2 1/2 inch hose and weighed around 4,000 pounds.

Engine Company No. 28

Served the 20th Ward from its engine house at the corner of Filbert and Elmer streets. No. 28 was ready to answer the fire call on November 3, 1899.

A second-size Metropolitan crane-neck engine built by the LaFrance Fire Engine Co. was assigned there. A three-horse affair, this engine weighed 8,300 pounds and was placed in service on May 25, 1900. The builder's serial number was 2719.

On June 14, 1900, a two-horse combination Champion chemical engine and hose wagon was delivered to No. 28. On board was a large deck turret, 60-gallon chemical tank and 1,000 feet of 2 ply Paragon Hose. It was manufactured by the Fire Extinguisher Manufacturing Co. and weighed 6,500 pounds.

Stationed in the same house, Chemical Engine Company No .2 used a four-wheel Champion double-tank chemical engine with two 60-gallon tanks.

It was placed in service on December 23, 1902, after completion at the Fire Extinguisher Manufacturing Co. plant. Two horses were required to pull its 4,500-pound weight.

Engine Company No. 29

Was located at Hamilton and Braddock avenues, activated on December 3, 1900.

A third-size double-pump Amoskeag engine served this company and was placed in service in June 1883. The plant erection number was 570. It weighed 6,400 pounds and needed two horses to transport it.

It was attended by a four-wheel hose carriage built by the William H. Leonhard Wagon Works of Pittsburgh and was delivered on December 3, 1900. Carried on board is 1,000 feet of size 2 1/2 inch hose. Two horses pulled the 3,600-pound wagon.

Engine Company No. 30.

Located in downtown Pittsburgh, this company resided at First Avenue below Smithfield Street. Originally, it was placed in service at the corner of Strawberry and Cherry alleys on September 1, 1900. They moved to their First Avenue

With esthetic grace and massive power, Engine Company 28's three matched browns take their charge into a left turn on this bright summer day. (Carnegie Library of Pittsburgh)

A study of facial expressions, a youthful audience gathers to watch the old "Hope" steamer of Allegheny City at work. No matter what the era, young and old alike have a fascination for fire engines at work or rest. (Carnegie Library of Pittsburgh)

address on November 27, 1900.

Engine equipment was a first-size American engine with a crane-neck and double pumps. Three horses pulled this engine, which was placed in service on September 4, 1900.

Hose carriage was the standard department prototype, this one built by the William H. Leonhard Co. and placed in the enginehouse for duty on September 4, 1900, two horses pulled it.

Engine Company No. 31

One of the newer companies, it was placed in service on November 1, 1904. Its engine house was located on Winterburn Street in the Garfield District of the city.

Equipment assigned to No. 31 was a four-wheel hose wagon drawn by two horses. The wagon carried 1,000 feet of 2 1/2 inch hose

Engine Company No. 32

This company was originally placed in service in an annex at the rear of Engine Company No. 3 on February 1, 1901. One year later, on February 1, 1902, the company moved into its present home on Eighth Avenue below Penn.

This company had the distinction of having the only self-propelled steam fire engine in the Pittsburgh fire service. It was a double-extra first-size Amoskeag unit, placed in service on February 1, 1901. Propulsion was furnished by steam taken from the engine's boiler. The engine had a total weight of 17,000 pounds and was manufacturer's serial number 756.

A Champion combination chemical and hose wagon traveled with this engine which was pulled by two horses. It was manufactured by the Fire Extinguisher Manufacturing Co. of Chicago and features one 60-gallon chemical tank and a turret deck nozzle. It was placed in service on June 23, 1900.

Engine Company No. 32s combination wagon, built by the Fire Extinguisher Manufacturing Company of Chicago, Illinois. The photo was taken on Eighth Street in the year 1900. (Carnegie Library of Pittsburgh)

Engine Company No. 33

Located on Eighth Street below Penn Avenue, Activation date for this company was June 13, 1906.

Its engine was an extra first size crane neck double-pump affair, manufactured by Amoskeag as its serial number 700.

In attendance was a Holloway combination chemical and hose wagon built by the American-LaFrance Fire Engine Co. of Elmira, New York. It was placed in service on December 1, 1906, and was equipped with a glacier nozzle and was drawn by two horses.

Engine Company No. 34

Located on Northumberland Street and Asbury Place, this company was activated on December 1, 1906.

This company used a third-size crane-neck double-pump engine of the Amoskeag nameplate. The engine weighed in at 7,200 pounds and required two horses to get it to the fire scene. It

Engine Company 32's self-propelled Amoskeag engine was in service only one month when this photograph was taken on March 1, 1901. The scene of the fire is the Exposition grounds. During the alarm, this engine was in continuous service for a record 30 hours straight. (Carnegie Library of Pittsburgh)

During the flood of March 1907, many city engine companies were pressed into service pumping flood waters out of downtown Pittsburgh streets. (Carnegie Library of Pittsburgh)

Engine Company No. 35

Was located in the 40th Ward at Tabor and Radcliffe Streets and began service December 7, 1906. This location was the former town hall, police station, and fire station of the Borough of Esplen. After extensive renovation, it was converted for use as a city fire enginehouse.

Equipment at this station was a James G. Weir and Sons four-wheeled hose carriage pulled by two horses. It was activated for service when the station was opened on December 7, 1906.

Truck Company "J"

On July 2, 1900, Truck Company "J" was ordered to service at its assigned engine house at number 49 South 14th Street.

Their equipment was a 50-foot Babcock type aerial hook-and-ladder truck built by the Fire Extinguished Manufacturing Co. of Chicago. Two horses drew this engine, which was put in service on July 2, 1900.

Chemical Company No. 1

Located in the uptown section at number 40 Tunnel Street, this company first saw service on November 21, 1892 on Liberty Avenue near Fourth Street, the rear of Engine House No. 1. They remained there until March 19, 1906, when the Wabash-Pittsburgh Terminal Railroad purchased that property. The following day, the company moved into their present Tunnel Street quarters. This site was also the location of the Bureau of Fire horse stables.

Their apparatus was a double 60 gallon tank Holloway chemical engine drawn by two horses. It was placed in service on November 21, 1892, the machine weighed in at 4,850 pounds.

Water Tower Company No. 1

Was situated on Smithfield Street near Second Avenue and was the property formerly occupied by the No. 2 Engine Company. Inservice date was January 7, 1900.

Its rig was a 65-foot Champion Water Tower, built by the Fire Extinguisher Co. and placed in service on January 7, 1900.

Some information exists that Engine Company No. 38 was in service at Lemington Avenue and Missouri Street on or about October 21, 1907, but no definite evidence could be found to support this claim.

North Side fire companies brought into the greater Pittsburgh system included 10 engine companies and 6 hose companies. The numeral 4 was added in front of the old North Side company number, making old Engine Company No. 1 now new Engine Company No. 41 and so on through Engine Company No. 49. One exception to this was Engine Company No. 10 being redesignated new Engine Company No. 50. This also permitted

was officially assigned to No. 34 in October 1885.

Assisting at fires was their James G. Weir combination chemical engine and hose wagon. It was delivered to No.34 on December 13, 1906. Built on the wagon chassis were two 35-gallon chemical tanks, two horses were assigned to pull it.

Truck Company "K" was located in the same structure with No. 34 and was a Seagrave city service truck built by the Seagrave Apparatus Co. of Columbus, Ohio. Ladder equipment included one 50-foot extension, one 40-foot extension, one 30-foot, one 25-foot, one 20-foot, one 16-foot extension, one 16-foot and one 12-foot. December 1, 1906, was its first day in the fire service. It was drawn by two horses.

the old Hose Companies No. 1 through 6 to be renumbered 51 through 56. Company No. 54 was converted to an engine company before the consolidation was finalized. North Side facilities numbered 16 engine houses with a total of 10 engine companies, 5 hook & ladder companies and 6 hose companies.

A roster of the new Pittsburgh fire bureau units listed the following:

Engine Company No. 41

This engine company was located at Martin and Corry streets in the old Hope Company enginehouse and was put into service in 1882. Equipment was a second-size Amoskeag engine which could deliver 550 gallons per minute. A new water tube boiler was installed in this unit by the American-LaFrance Company in 1907.

Engine Company No. 42

Was located at Madison Avenue near Pike Street and was activated in 1894. Its engine was a second-size Amoskeag pumper with a capacity of 700 gallons per minute. This engine was overhauled in 1907.

Engine Company No. 43

In 1906, this company was put into service at Arch and Jackson streets. Serving this company was a third-size Metropolitan engine with a pumping capacity of 650 gallons per minute.

Engine Company No. 44

Manhattan and Franklin streets was the headquarters for this engine company, which provided service with a second-size Metropolitan engine, which delivered 700 gallons of water per minute; it was activated in 1906.

Engine Company No. 45

The No. 45 Company was situated on Preble Avenue near Kerr Street, with a second-size Amoskeag engine that could produce 550 gallons of water per minute. Put into service in 1874, the engine was refurbished in 1908.

Engine Company No. 46

Had its enginehouse located on Sandusky Street near Park Way. Its engine was a first-size Amoskeag engine, which had been built in 1889. It was rebuilt with a new boiler in 1899. Pumping capacity was 700 gallons per minute. Sharing quarters with No. 46 was Truck Company "L."

Engine Company No. 47

Fulton and Lincoln Avenue was the address for this company. The premises were originally home to the Eureka Fire Company. A second-size Silsby engine with a 700 gallon-per-minute capacity served this company. The engine was first put into service in 1884. Supporting this company was Truck Company "M."

Engine Company No. 48

This unit was headquartered on River Avenue near Pike Street and had as its engine a second-size Silsby type, which was put into service in 1886. The pump output for this engine was 700 gallons per minute. A large limestone insert on the front of the structure identified it as the former home of the Ellsworth Volunteer Fire Company.

Engine Company No. 49

Was located along busy Spring Garden Avenue near Lager Street with a second-size Amoskeag pumper, which could produce 550 gallons per minute. The unit was activated in 1874 and received a new boiler in 1888. The classic structure was formerly home to the Spring Garden Engine Company.

Engine Company No. 50

Had its enginehouse erected on Lafayette Avenue near Federal Street and was served by a third-size Metropolitan engine activated for service in 1902. The engine could deliver 450 gallons of water per minute. Sharing space with this company was Truck Company "N."

Hose Company No. 51

Was located at Branch and Lowrie streets along with Truck Company "P."

Hose Company No. 52

Along with Truck Company "O," was located at Shady Avenue and Dixon Street.

Hose Company No. 53

Rhine Street, Spring Hill, was the headquarters for this hose company.

Bureau of Fire Fuel Wagon Number 4 is making a delivery in support of Engine 19's work. Among the details in this scene is the whip, set in its socket beside the engine driver's seat. (Carnegie Library of Pittsburgh)

Newly activated Engine Company No.60, Fallowfield Avenue in Beechview, has turned out their 1912 American-LaFrance chemical and hose truck early on this autumn morning, even beating the milkman as evidenced by the empty bottle waiting on the house porch. This truck was the first motorized piece of fire equipment purchased by the City of Pittsburgh.
(Willie E. Patterson, Photographer, Collection of Robert W. Lewis)

Engine Company No. 54

Located on Perrysville Avenue at Broadway, this company was placed in service with a Nott fourth-size engine which delivered 450 gallons of water per minute.

Hose Company No. 55

Arthur Street in the 27th Ward was the location of this hose company.

Hose Company No. 56

A Spring Hill hose brigade, this company was located along South Side Avenue at Luella Street.

Three engines were held in reserve for the North Side district, relics from the early Allegheny Fire Department. The best of the lot was a second-size Amoskeag built in 1877, and re-boilered in 1886. It could produce 550 gallons at the nozzle. A third-size Amoskeag, vintage 1869 and a first-size Silsby built in 1889, were worn out and questionable for service other than as back-up or for auxiliary pumping.

A total of 5,936 fire hydrants were in active service during 1908. This included the city of Pittsburgh proper, North Side, Beltzhoover, Esplen, Sheraden, and North Homestead.

Except for Engine Company No. 32, all bureau engines, trucks and wheeled equipment were drawn by horses. The lone piece of self-propelled apparatus was No. 32's steam-powered double extra first-size Amoskeag fire engine. Horse power was still king in the Pittsburgh fire department, with a grand total of 209 four-legged employees serving the bureau. Internal combustion was still a few years off from making its debut in the city firehouses, but changes were coming. A milestone was reached in 1908 when the last steam-powered self-propelled Amoskeag engine left the factory assembly shop. Like the old faithful horses, steam power was headed into the sunset.

Gasoline-powered firefighting equipment was already in limited use in other cities for several years now. Internal-combustion engines were first used to pump the water to the hoses, while horses still pulled the engine itself. The first commercially successful motorized piece of fire equipment utilizing petroleum fuel was a combination hose and chemical wagon manufactured by the American-LaFrance Fire Engine Co. in 1904. Other builders, including Seagrave, Howe, and Webb were experimenting with their own designs and testing prototypes in field situations.

Some of the automobile manufacturers who were dabbling in fire engine design included Knox, Pope, Kissel, White, and Reo. In 1906, the Waterous Engine Works attached a 300 gallon-per-minute gasoline-powered pump to the rear of an automobile chassis, making it the first American fire unit to use internal combustion to both drive the machine and power the pumps. A new company formed in 1908, Ahrens-Fox, would use the marvel of gasoline to fuel its highly advanced pump designs and to power its massive and powerful multi-cylinder truck engines. At Pittsburgh, Chief Humphreys and his staff kept abreast of these and other advances in the art of building firefighting equipment.

In early 1909, Chief Miles Humphreys requested authority from the mayor and council to create a standard design for the city's first gasoline-powered fire truck. The prototype was a combination truck of American-LaFrance manufacture, and was delivered that summer. After the city fire bureau made some modifications, including the addition of two hose reels, the new engine was tested at Silver Lake in the east end section of the city.

During this period several other fire companies were activated by the Bureau of Fire. No. 36 took up station at Stanton Avenue and Hawthorne, and No. 37 was assigned to duty at Rebecca and Columbo streets. The West End section of town welcomed Engine Company No. 39 into their headquarters at Lorenz Avenue and Steuben Street. No. 40 firemen reported to their engine house at Chartiers Avenue and Citadel Street, while No. 57 opened for service on Brookline Boulevard at Crawford Street, and the No. 58 company moved into space on Allequippa Street. Completing the numerical sequence was Engine Company No. 38 which had previously been assigned to its Lemington Avenue and Missouri Street location.

Liberty Avenue continued to be the scene of noteworthy fires, with September 1909 providing one event that would not soon be forgotten. Between Ninth and Wood streets stood the six-story building at number 819, whose main tenant was the Sterling Stamp Co. Adjacent to it, and under construction, was the new Second National Bank building. Around a quarter past four in the afternoon, smoke and then fire rose up from the top of the stamp company structure. The probable cause was a heated construction rivet that missed the catcher's bucket when thrown from the forge and fell on top of Sterling's roof.

At about this same time, a nearby theater completed their matinee performance and emptied its patrons out onto the street. Timing was perfect, there would be an unscheduled second show today. When the downtown engine companies pulled up and began to dismount, sidewalks filled with curious citizens anxious to see the firemen start their attack. Mounted police began to press the crowds back from the curbs, while foot patrolmen cordoned off the avenue and connecting streets. The raging fire wasted little time in burning its way down to the street level. In what seemed like a matter of minutes, fire was shooting from every window, heat bursting the glass panes outward to relieve the scorching pressure.

With three alarms activated, multiple hoses played water streams on the burning hulk. Firefighters raised their ladders and brought up hoses to reach the upper floors and the building's interior. Overhead wires in the street and along the curb made it difficult for firemen to erect ladders in a safe manner. The No. 1 Water Tower

Company had tough going trying to position its equipment properly. Truck Company No.2's 60-foot ladder became overstressed under the weight of men and equipment and buckled in against the building face. As onlookers gasped, the remaining upper section, with four men aboard, began to bend backward and twist in the same motion. Without hesitation, the four descended the swaying ladder, leaving their hoses behind. Fortunately, all bureau fire ladders were reinforced with steel rods along their entire length. This factor alone probably saved the four firemen from severe injury, or possibly, death. Within 90 minutes only the smoking black shell of the building was still standing. Damages to the structure and its contents was set at $50,000.

Growing in the number of new companies and additional men added to the force, the 1910 Bureau of Fire rostered 82 companies of which there were 45 engine companies, 14 hose companies, 2 chemical companies, 20 hook-and-ladder companies and one water tower. Personnel totaled 910, of which there were 847 uniformed men and 63 non-uniformed men.

By the year 1911, Pittsburgh's rate of land expansion had grown again. Over the past four years, two more boroughs were brought into the city family, covering an area of just over three square miles. Recent municipal additions included West Liberty and Beechview Boroughs. The entire area was gaining stature as one of America's most progressive metropolitan cities. Consider that Pittsburgh was home to 565,000 people, had 500 miles of paved streets, 13 public parks and over 450 churches and missions. Property valuations were estimated at $700 million.

Fiscal operations of the fire bureau passed the two million dollar mark during 1911, of which $1,152,973.50 went for maintenance. Yearly salaries for Chief Humphreys and his deputy chiefs

Bureau of Fire personnel test the pumping capabilities of a new gasoline powered machine. The year is 1909 and the location is Silver Lake in the City's East End section. The large stone viaduct carries the Brilliant Division of the Pennsylvania Railroad. (Carnegie Library of Pittsburgh)

This five man ladder truck crew proudly displays their equipment and the large, beautiful two-horse team that pulled it. Take special note of the size of the Tillerman's steering wheel. The days of horse-drawn equipment were declining by the mid-1900s. (Carnegie Library of Pittsburgh)

were $4,000 and $3,000 respectively. The 8 district chiefs earned $2,000 per year. captain and lieutenant line officers made $100 and $97.50 each month. The 46 engineers and assistant engineers took home pay envelopes with $100 and $90. Department drivers were a valuable commodity in the bureau, and their $95 monthly wage reflected this. At the bottom of the pay scale were 432 hosemen and laddermen who, for first, second and third grade positions, claimed $90, $85 and $80 a month. For all full time regular employees, other benefits included 14 days annual vacation, one day off after every five worked and 3 hours allotted per day for meals.

With each passing day, the Pittsburgh Bureau of Fire had more citizens and property to protect. Chief Humphreys realized that his forces were fighting fires in 20th century times with 19th century equipment. Firemen were faced with covering territories over and above their physical and equipmental capabilities. The gap between motorized firefighting equipment technology and the horse-and-buggy was growing wider each day. Humphreys petitioned the mayor and councils to appropriate funds to purchase new machines with modern features. His master plan was to begin replacing elderly and obsolete horse drawn fire engines with up-to-the-minute trucks equipped with motorized pumping and propulsion capabilities. With the city fathers' blessing, a motorized American-LaFrance truck was ordered in the summer of 1911.

Illustrating this urgent need was the Iron City Grain Elevator fire on November 8, 1911. The building, an old frame affair, was located along the 1300 block of West Carson Street on the South Side. A passing eastbound train's errant firebox coal fell on the bare, dry building frame, setting it on fire. The flames grew unchecked until the next train came along at 3 p.m. A startled railroad crew stopped its freight train and ran up to Carson Street to report it. Due to the 100 by 150 foot size of the building, Iron City employees were unaware that the rear of their building was burning.

They evacuated the premises none too soon for the fire was spreading fast. Firemen were kept back from the flames and intense heat as the tinder dry building walls were reduced to ashes in no time. Across the way, stables of the Mutual Union Brewing Co. were evacuated as a precaution. Forty very nervous horses, with smoke-filled nostrils, were led away to safety.

Suddenly, a front wall caved in, releasing tons of stored grain out onto busy Carson Street, closing it off from traffic. Trolley service was halted due to the mountain of kernels, forcing patrons to walk the remaining distance to their destinations. By the time the second alarms answered, the fire had burned itself down to the ground. Ruined by the fire were 75,000 bushels of grain that, when added to the fire ravaged value of the four story building, totaled $150,000 in losses. Four nearby homes were destroyed when airborne sparks and embers fell on them. During the blaze, Captain Matthew Reilly's horse bolted away when the pole he was tied to caught fire. When the fire scene was secured, the No. 30 Engine Company foreman was seen running down Carson Street to find the animal.

Fire engine technology was advancing quickly, growing by leaps and bounds including Seagrave's first internal-combustion-powered pumper followed by the first such vehicle of Mack's manufacture, also a pumper. Perhaps the most important design advance was Ahrens-Fox's piston pump that was mounted in the front of the engine truck's hood. Pittsburgh finally embraced the new wonder of internal combustion on November 14, 1911, when American-LaFrance delivered the first motorized truck. The new fangled machine was assigned to the six men of the No. 60 Engine Company located on Fallowfield Avenue in the Beechview section of the city. Commissioning took place on February 2, 1912. This new apparatus was a Type 10 combination hose and 40-gallon chemical tank truck built by American-LaFrance as factory serial number 87. Its power plant was a gasoline-powered in-line four-cylinder, 75-horsepower engine with chain-driven rear wheels.

Shortly after, the city purchased a 1912 Webb motorized chemical and hose wagon for Engine Company No. 31. But American-LaFrance's Type 10 design proved to be the most successful, especially for outlying Pittsburgh neighborhoods, where the terrain was steep and great distances had to be traveled to and from fire calls. In August 1912, the Bureau of Fire contracted with LaFrance for the delivery of three more Type 10 trucks, they being serial numbers 168, 169 and 170.

Just as the new was coming in, the old was going out. This year brought the end of production for Ahrens-Fox steam-powered engines. The year also served as the high-water mark for the number of horses in the Pittsburgh fire bureau. On December 31, 1912, they numbered 314, next year, only 285 would be in harness. In addition to the comparisons of horse power versus horsepower, there was an economic consideration as well. The costs of feeding and caring for the needs of a three-horse team came to approximately $650 a year. On the other hand, a gasoline powered fire engine could be maintained for about $100 a year including the fuel.

Although the use of horses in the fire bureau was on the decline, they remained the prime mover for most other Pittsburgh businesses. Both would meet each other at the Adams Express Company fire of December 2, 1912. The three-story brick building at the corner of 13th and Liberty fell victim to a fire that originated in the company clock at the front of the building. Gas was used to illuminate the clock face and this was thought to have been the culprit. Although the building shell was masonry, the interior was full of combustibles, notably hay and feed for the company's horses. These products ignited, causing a rapid chain reaction of flame.

When the first fire companies arrived, they could hear the frantic calls of horses inside, the acrid smoke and roaring fire swirling around them. Their frightened cries struck terror in the firemens' own horses, which quickly became nervous and jittery, sensing the plight of their trapped brothers and sisters. Fast action by the firemen and stable hands saved the day. While their mates laid hose and started to pour on the water, other firemen, in full regalia, went into the burning building and led out all 70 horses stabled on the second and third floors to safety. All would live to work another day.

In late 1912 and early 1913, nine Christie two-wheel tractors were purchased by the city to replace horse power on a like number of steam-boilered engines. Five new Knox-Martin three-wheel tricycle-type tractors were ordered as replacements for horse teams on three aerial ladder trucks and two hook-and-ladder trucks, generically known as city service trucks. The Knox units proved to be an immediate failure. When making turns, their front wheel came in contact with city street car tracks, tipping the trucks over.

Minor adjustments in the size of Bureau of Fire staff brought their 1913 complement to 811. Chiefs, deputies and assistants remained at 8 without any salary change. There was one less captain in the department, but the monthly wage was raised to $125. The 75 lieutenants now made $107.50 each month, up from $97.50. Engineers and assistant engineers still totaled 46 each, but their pay was raised $10 per month each to $110 and $100. Drivers also received a $10 monthly raise in salary and added one more man to their rolls. Hose and ladder men, the largest single category, listed 435 men. Their wages remained the same. Rounding out the list were 8 fuel wagon drivers who were paid $105 each month. Employee benefits remained the same.

Fire Bureau management continued to tinker with equipment design, adding 10 new Knox

On Friday, October 15, 1915, a fire at the Union Paper Box Company on Sandusky Street, North Side, claimed the lives of a dozen young women employees. Streetcars were halted in both directions as city firemen attempted to extinguish the blaze. (UPI/Corbis Bettmann)

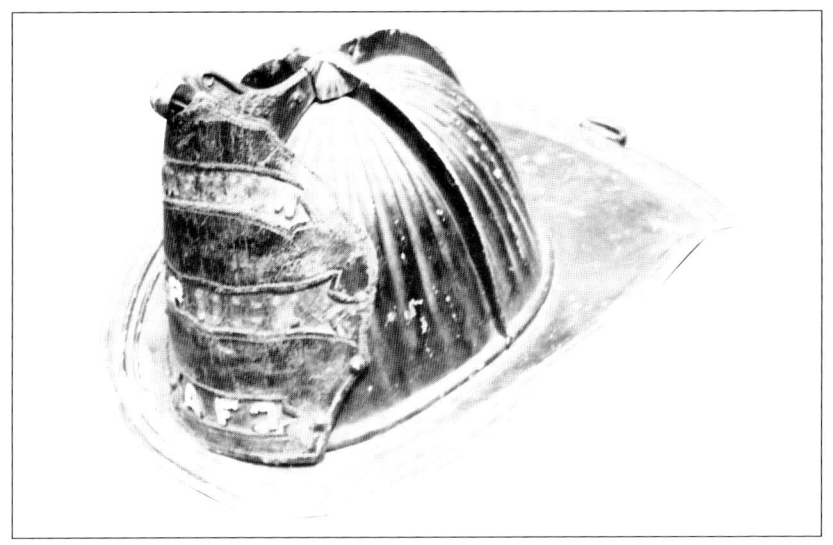

above, **Showing the battle scars of many fires fought, this fireman's helmet from Allegheny City could probably speak volumes about the life and times of early firefighters.** (Collection of Richard L. Linder)

Captain George King blew this trumpet loud and long to clear a path for his engine to get to the fire scene. It was the tradition, carried on each company roster, for its Captain to be listed "On the trumpet" (Collection of Richard L. Linder)

chemical and hose trucks to the force in 1913. These were assigned to engine companies 4, 5, 6, 10, 14, 17, 25, 26, 34, and 36. Also known as chemical and hose cars, they were powered by four cylinder inline gasoline-fueled engines. These units were never as popular as the American-LaFrance trucks, which were quickly becoming the standard Pittsburgh department prototype.

The year 1913 was a transition year for the top management of Pittsburgh's Bureau of Fire. Assistant Chief William Coates was promoted to chief of the bureau, a post he would hold until the personnel realignment made by newly elected Mayor Joseph G. Armstrong in 1914.

Meanwhile, old standby equipment designs continued to fade from use as new inventions appeared in their places. Motorized chemical engines became a staple in most every fire department, including Pittsburgh's. Horse-drawn steamers began to decline in popularity due to limited capacity and operating range, especially in suburban neighborhoods. Although they remained in use for a number of years after, their zenith years were past and they were scrapped when mechanical or other maintenance problems could not be economically solved. One new item in which every department was vitally interested was the pressurized booster tank invented by Ahrens-Fox in 1913.

Pittsburgh's fire service continued to place confidence in American-LaFrance products, with an order for 10 additional Model 10 combination chemical and hose trucks during the period December 1913 to January 1914. The Model 10 was a very popular design that had changed little since its introduction. Only a slightly different front fender configuration distinguished it from its older brothers. These new trucks were built with factory serial numbers 442 to 451.

The year 1914 brought still more changes and refinements. American-LaFrance rolled out its last steam-powered fire engine from the assembly plant floor. That same year, the company produced its first motorized water tower. Years later, in 1938, American La-France would also gain the distinction of producing the last American-made water tower, a 65-foot long unit. Other companies were turning out new equipment to meet the increasing demand for internal-combustion powered apparatus. Manufacturers like Maxim, Webb, and Nott, among others, continued to expand and build on the horseless carriage technology. It was in this year of 1914 that the shrouded, front-mounted pump mechanism of Ahrens-Fox appeared with its highly polished, spherical-shaped air chamber, which became the visual trademark by which all future generations of its equipment would instantly be known.

Pittsburgh's Bureau of Fire was now three years into its equipment modernization and replacement program. New American-LaFrance chemical engines and hose trucks were assigned to strategic locations throughout the city. Older equipment was downgraded to lesser duty stations in town, sold off to other communities, or cannibalized for needed parts. Some of these old veterans had been in service for more than 40 years. Changes were affecting the horses as well. Some of them found their days and nights less demanding, while others went to homes on farms. Some were sold, but most spent their remaining years doing easier work.

Replacing the four-legged power were 15 new American-LaFrance Type 31 two-wheeled tractors delivered to the bureau during the months of May to August 1915. Each was powered by a four-cylinder engine that developed 75-horsepower. Maximum speed was 25 miles-per-hour, mainly because of low gear ratios needed to pull the old steam-boilered engines. The gearbox was a three-speed type.

Two were assigned to pull the LaFrance steamers of companies 4 and 8. One each was fitted to a LaFrance Metropolitan horse drawn steamer, a Babcock horse drawn aerial ladder truck and a 1909 Seagrave horse drawn hook and ladder truck. Two found their homes with elderly Amoskeag steam engines, formerly transported by horse power. The remaining eight were delivered with new American-LaFrance Metropolitan steam boilered engines. Engine companies 46 and 47 received 900 gallon per minute first size types. Engine companies 10, 11, 12, 42, 45, and 48 were assigned the second size 700 gallon per minute engines. These were the last steam fire engines purchased by the Pittsburgh Bureau of Fire.

Equipment drivers also faced changes. Gone were the personal bonds that matched teams of horses and drivers, replaced by an impersonal slender metal hood with 40 to 75 eager horses underneath. Hands that were worn and shaped by years of working the leathers now held a hard round wheel to steer the engine. The driver didn't have to exhort his team on to the fire, only sit there, silent, letting mechanical power answer to his commands of feet and hands.

Even the lives of fire bureau chiefs began to change. On March 1, 1915, a two-passenger Overland roadster automobile, Model 81R, serial number 10194, was placed in service at District No. 6 for the use of Chief Fred Beckett. His district commanded Engine Companies 8, 16, 27, 29, 34, and 38, plus Truck Companies 2, 9, 11, and 19. Returned to the Tunnel Street stables was a veteran two-seat horse-and-buggy rig, made obsolete by the internal combustion engine.

Another era of Pittsburgh firefighting was coming to an end, hastened by technological developments, progress, and the war in Europe. By the time the Armistice was sounded three years later, steam-powered fire engines and the magnificent horses that pulled them would be a fading memory. Even the firehouse dog was a vanishing breed. In six years, every fire company would have at least one piece of motorized work equipment.

Now it was farewell to stables, hay, and sawdust; to leather harness, and cracking whips; good-bye to tin hats, trumpets, and carbide lanterns; adieu to old traditions, spittoons, endless storytelling and firehouse life; a last look at those graceful, galloping horse teams driven by iron men on smoking engines with steam-powered whistles, polished brass, and painted wheels.

The first 125 years of Pittsburgh fire department life was committed to history. ☆

Posed outside the Second Presbyterian Church, the Number 19's Deluge Wagon and relaxed crew wait to step off, possibly for a parade. The well cared for three-horse team are at the ready. (Carnegie Library of Pittsburgh)

In 1907 the Pittsburgh Bureau of Fire hosted a gala celebration for the 28th annual meeting of the fire brotherhood. Number 19's handsome engine house is decorated with flags and buntings to welcome the visiting firemen. (Willie E. Patterson, Photographer, Collection of Robert W. Lewis)

Appendix One - Equipment

Steam-powered pumping engines of the City of Pittsburgh Bureau of Fire, circa 1908.

Engine Company No. 1 second-size Amoskeag engine, 4 1/4 inch diameter pumps, cylinder diameter 8 inches by 8 inch stroke, height 9 feet, width 6 feet, length 24 feet 3 inches. Builder's #751.

Engine Company No .2 first-size Amoskeag engine, 4 3/4 inch diameter pumps, cylinder diameter 9 inches by 8 inch stroke, height 9 feet 2 inches, width 6 feet 2 inches, length 24 feet 6 inches. Builder's #723.

Engine Company No. 3 first-size Amoskeag engine, 4 3/4 inch diameter pumps, cylinder diameter 9 inches by 8 inch stroke, height 9 feet 2 inches, width 6 feet 2 inches, length 24 feet 6 inches. Builder #750.

Engine Company No. 4 second-size LaFrance engine, 4 3/4 inch diameter pumps, cylinder diameter 7 3/4 inches by 9 inch stroke.

Engine Company No. 5 third-size Amoskeag engine, 3 7/8 inch diameter pumps, cylinder diameter 6 1/8 inches by 8 inch stroke, height 8 feet 10 inches, width 6 feet 2 inches, length 24 feet. Builder's #601.

Engine Company No. 6 third-size Amoskeag engine, 4 inch diameter pumps, cylinder diameter 6 inches by 8 inch stroke, height 8 feet 10 inches, width 6 feet, length 24 feet. Builder's #617.

Engine Company No. 7 second-size Amoskeag engine, 4 3/4 inch diameter pumps, cylinder diameter 9 inches by 8 inch stroke, height 9 feet, width 6 feet, length 24 feet 3 inches. Builder's #452.

Engine Company No. 8 second-size LaFrance engine, 4 1/2 inch diameter pumps, cylinder diameter 7 1/2 inches by 8 inch stroke. Builder's #313.

Engine Company No. 9 second-size Amoskeag engine, 4 1/8 inch diameter pumps, cylinder diameter 8 inches by 9 inch stroke, height 9 feet, width 6 feet, length 24 feet 3 inches.

Engine Company No. 10 second-size Amoskeag engine, 4 1/4 inch diameter pumps, cylinder diameter 6 7/8 inches by 8 inch stroke, height 9 feet, width 6 feet, length 24 feet 3 inches.

Engine Company No. 11 second-size Amoskeag engine, 4 1/4 inch diameter pumps, cylinder diameter 6 7/8 inches by 8 inch stroke, height 9 feet, width 6 feet, length 24 feet 3 inches. Builder's #262.

Engine Company No. 12 third-size Amoskeag engine, 4 3/8 inch diameter pumps, cylinder diameter 8 inches by 9 inch stroke, height 8 feet 10 inches, width 6 feet, length 24 feet.

Engine Company No. 13 fourth-size Amoskeag engine, 4 3/8 inch diameter pumps, cylinder diameter 8 inches by 9 inch stroke, height 8 feet 3 inches, width 5 feet 10 inches, length 20 feet 3 inches.

Engine Company No. 14 fourth-size Amoskeag engine, 4 1/2 inch diameter pumps, cylinder diameter 7 1/2 inches by 8 inch stroke, height 8 feet 3 inches, width 5 feet 10 inches, length 20 feet 3 inches.

Engine Company No. 15 first-size Amoskeag engine, 4 3/4 inch diameter pumps, cylinder diameter 9 inches by 8 inch stroke, height 9 feet 3 inches, width 6 feet 2 inches, length 24 feet 6 inches. Builder's #698.

Engine Company No. 16 second-size Manning engine, 4 1/2 inch diameter pumps, cylinder diameter 7 inches by 8 inch stroke. Builder's #114.

Engine Company No. 17 fourth-size Amoskeag engine, 4 inch diameter pumps, cylinder diameter 8 inches by 9 inch stroke, height 8 feet 3 inches, width 5 feet 10 inches, length 20 feet 3 inches. Builder's #403.

Engine Company No. 18 first-size Amoskeag engine, 4 3/4 inch diameter pumps, cylinder diameter 9 inches by 8 inch stroke, height 9 feet 2 inches, width 6 feet 2 inches, length 24 feet 6 inches.

Engine Company No. 19 Extra first-size American engine, 5 3/4 inch diameter pumps, cylinder diameter 9 1/2 inches by 9 inch stroke.

Engine Company No. 21 fourth-size Amoskeag engine, 4 3/8 inch diameter pumps, cylinder diameter 8 inches by 9 inch stroke, height 8 feet 3 inches, width 5 feet 10 inches, length 20 feet 3 inches. Builder's #477.

Engine Company No. 24 third-size Amoskeag engine, 4 inch diameter pumps, cylinder diameter 6 inches by 8 inch stroke, height 8 feet 10 inches, width 6 feet, length 24 feet.

Engine Company No. 25 second-size LaFrance engine, 4 5/8 inch diameter pumps, cylinder diameter 7 3/4 inches by 9 inch stroke. Builder's #425.

Engine Company No. 26 fifth-size American engine, 3 5/8 inch diameter pumps, cylinder diameter 5 inches by 6 inch stroke.

Engine Company No. 27 third-size Amoskeag engine, 3 3/4 inch diameter pumps, cylinder diameter 6 inches by 8 inch stroke, height 8 feet 10 inches, width 6 feet, length 24feet. Builder's #582.

Engine Company No. 28 second-size Metropolitan engine, 4 3/4 inch diameter pumps, cylinder diameter 8 inches by 9 inch stroke. Builder's #2719.

Engine Company No. 29 third-size Amoskeag engine, 3 3/4 inch diameter pumps, cylinder diameter 6 inches by 8 inch stroke, height 8 feet 10 inches, width 6 feet, length 24 feet. Builder's #570.

Engine Company No. 30 first-size American engine, 5 1/4 inch diameter pumps, cylinder diameter 8 inches by 10 inch stroke.

Engine Company No. 32 Double extra first-size Amoskeag engine, 5 3/4 inch diameter pumps, cylinder diameter 9 1/2 inches by 8 inch stroke, height 10 feet, width 7 feet 3 inches, length 16 feet 6 inches. Builder's #756.

Engine Company No. 33 Extra first-size Amoskeag engine, 5 3/4 inch diameter pumps, cylinder diameter 9 1/2 inches by 8 inch stroke, height 10 feet, width 6 feet 5 inches, length 24 feet 9 inches. Builder's #700.

Engine Company No. 34 3rd third-size Amoskeag engine, 4 inch diameter pumps, cylinder diameter 6 inches by 8 inch stroke, height 8 feet 10 inches, width 6 feet, length 24 feet. Builder's #607.

Amoskeag Steam Fire Engines, manufactured by the Manchester Locomotive Works of Manchester, New Hampshire, and delivered to the fire departments of Pittsburgh and Allegheny City. Lettering listed was on the engine when it was delivered. On many units, lettering changed several times during their years of service.

Builder's #57, Second Size, Single Pump, Harp Tank style, lettered "Neptune," delivered to Pittsburgh, Pa., September 1862.

Builder's #73, First Size, Double Pump, Round Tank style, lettered "Hope," delivered to Allegheny City, Pa., July 1863.

Builder's #78, Second Size, Single Pump, Harp Tank style, lettered "Relief No. 9," delivered to Pittsburgh, Pa., December 1863.

Builder's #87, First Size, Double Pump, Round Tank, lettered "General Grant", delivered to Allegheny City, Pa., April 1864.

Builder's #114, Second Size, Single Pump, Harp Tank style, lettered "Niagara," delivered to Pittsburgh, Pa., August 1865.

Builder's #219, Second Size, Double Pump, Straight Frame style, lettered "Vigilant 6," delivered to Pittsburgh, Pa., February 1867.

Builder's #262, Second Size, Double Pump, Straight Frame style, lettered "Duquesne," delivered to Pittsburgh, Pa., December 1867.

Builder's #278, Second Size, Double Pump, Straight Frame style, lettered "Independence," delivered to Pittsburgh, Pa., April 1868.

Builder's #309, Second Size, Double Pump, Straight Frame style, lettered "Friendship 3," delivered to Allegheny City, Pa., July 1869.

Builder's #366, Second Size, Double Pump, Crane Neck Frame style, lettered "Good Will 4," delivered to Allegheny City, Pa., June 1871.

Builder's #403, Third Size, Single Pump, Harp Tank style, lettered "Wm. Phillips 5," delivered to Pittsburgh, Pa., February 1872.

Builder's #450, Third Size, Single Pump, Harp Tank style, lettered "J. McC. Creighton 8," delivered to Pittsburgh, Pa., August 1873.

Builder's #451, Third Size, Single Pump, Harp Tank style, lettered "Jas. McAuley 11," delivered to Pittsburgh, Pa., August 1873.

Builder's #452, Second Size, Double Pump, Crane Neck Frame style, lettered "Wm. Lyons 12," delivered to Pittsburgh, Pa., July 1873.

Builder's #453, Second Size, Double Pump, Crane Neck Frame style, lettered "Henry Hays 13," delivered to Pittsburgh, Pa., June 1873.

Builder's #477, Third Size, Single Pump, Harp Tank style, lettered "Thomas Rees 10," delivered to Pittsburgh, Pa., July 1874.

Builder's #488, Second Size, Double Pump, Crane Neck Frame style, lettered "Lincoln 5," delivered to Allegheny City, Pa., September 1874.

Builder's #489, Second Size, Double Pump, Crane Neck Frame style, lettered "Columbia 6," delivered to Allegheny City, Pa., December 1874.

Builder's #512, Second Size, Double Pump, Crane Neck Frame style, lettered "General Grant 2," delivered to Allegheny City, Pa., April 1877.

Builder's #566, Second Size, Double Pump, Crane Neck Frame style, lettered "Hope," delivered to Allegheny City, Pa., June 1882.

Builder's #570, Third Size, Double Pump, Crane Neck Frame style, lettered "Engine No. 4," delivered to Pittsburgh, Pa., October 1882.

Builder's #582, Third Size, Double Pump, Crane Neck Frame style, lettered "Engine No. 6," Delivered to Pittsburgh, Pa., June 1883.

Builder #601, Third Size, Double pump, Crane Neck Frame style, Lettered "Engine No. 5," delivered to Pittsburgh, Pa., August 1884.

Builder's #607, Third Size, Double Pump, Crane Neck Frame style, lettered "Benjamin Darlington, Engine Company No. 14," delivered to Pittsburgh, Pa., October 1885.

Builder's #608, Third Size, Double Pump, Crane Neck Frame style, lettered "Robert C. Ellliott, Engine Company No. 1," delivered to Pittsburgh, Pa., May 1886.

Builder's #611, First Size, Double Pump, Crane Neck Frame style, lettered "Engine Company No. 2," delivered to Pittsburgh, Pa., March 1889.

Builder's #617, Third Size, Double Pump, Crane Neck Frame style, lettered "A.F. Keating, Engine Company No. 15," delivered to Pittsburgh, Pa., May 1886.

Builder's #618, Third Size, Double Pump, Crane Neck Frame style, lettered "Walter P, Hansell, Engine Company No. 12," delivered to Pittsburgh, Pa., May 1886.

Builder's #646, First Size, Double Pump, Crane Neck Frame style, lettered "Engine Company No. 3," delivered to Pittsburgh, Pa., August 1889.

Builder's #698, First Size, Double pump, Crane Neck Frame style, lettered "Engine Company No. 2," delivered to Pittsburgh, Pa., September 1893.

Builder's #700, First Size, Double Pump, Crane Neck Frame style, lettered "Engine Company No. 19," delivered to Pittsburgh, Pa., September 1893.

Builder's #723, First Size, Double Pump, Crane Neck Frame style, lettered "Engine Company No. 2," delivered to Pittsburgh, Pa., September 1896.

Builder's #724, Fourth Size, Double Pump, Crane Neck Frame style, lettered "Engine Company No. 26," delivered to Pittsburgh, Pa., September 1896.

Builder's #750, First Size, Double Pump, Crane Neck Frame style, not Lettered, Delivered to Pittsburgh, Pa., April 1900.

Builder's #751, Second Size, Double Pump, Crane Neck Frame style, Not lettered, delivered to Pittsburgh, Pa., April 1900.

Builder's #756, Double Extra First Size, Double Pump, Crane Neck Frame style, self-propelled, not lettered, delivered to Pittsburgh, Pa., December 1900.

Builder's #792, Second Size, Double Pump, Crane Neck Frame style, not lettered, delivered to Pittsburgh, Pa., 1906.

Builder's #793, Second Size, Double Pump, Crane Neck Frame style, not lettered, delivered to Pittsburgh, Pa., 1906.

Builder's #795, Second Size, Double Pump, Crane neck Frame style, not lettered, delivered to Pittsburgh, Pa., 1906.

Bibliography

Those wishing to read additional material on Pittsburgh's municipal history, including the fire department, and firefighting in general will find these titles of interest:

History of Pittsburgh
Killikelly, Sarah H., B.C. & Gordon Montgomery Co., 1906

Pittsburgh's Professional Firefighters
W.G. Johnston & Co., 1974

Pittsburgh & Her People, Volumes 1 through 4
Boucher, John Newton, Lewis Publishing Co., 1908

Fire Engines in North America
Buff, Sheila, The Wellfleet Press, 1991

History of Allegheny County, Volumes 1 and 2
A. Warner & Co., Publishers, 1889

Fire Engines, Fire Fighters
Ditzel, Paul C., Crown Publishers Inc., 1976

Select and Common Council Records, City of Pittsburgh,
Pennsylvania Historical Society of Western Pennsylvania Archives

Our Firemen
Charles T. Dawson Editor, 1889

History of the Allegheny Fire Department
John P. Shea, Publisher
1894, 1903, 1907

Memory's Milestones
Smith, Percy F., 1918

Notable Men of Pittsburgh and Vicinity
Smith, Percy F., Pittsburgh Printing Co., 1901

All Sorts of Pittsburghers
Burgoyne Arthur G., The Leader All Sorts Co., 1892.

Official Records
Pittsburgh Fire Department, 1870-1887
Pittsburgh Bureau of Fire, 1888-1915

Pittsburgh Today, Volumes 1 through 4
American Historical Society, Incorporated, 1931

Newspaper files of the
Pittsburgh Gazette Times
Pittsburgh Chronicle Telegraph
Pittsburgh Post
Pittsburgh Leader
Pittsburgh Daily Morning Post

Where's The Fire?, American Firefighters in picture postcards, circa 1910
Stein, Geoffrey N., The Vestal Press, Ltd., 1992

Annual reports:
City of Pittsburgh, Department of Public Safety
Bureau of Fire
Firemen's Disability Board
Board of Fire Escapes

American Fire Marks
The Insurance Company of North America, 1933

The American Fire Engine
Halberstadt, Hans, Motorbooks International, 1993

About the author . . .

Howard Worley's writing career began in 1961 when he created *The Modulator,* a monthly publication of the Five-Eleven Radio Club, Inc., of Pittsburgh, Pennsylvania. He served as its first editor, wrote articles, prepared art work, and pasted up the pre-press masters.

Next, he joined the Citizen Band Radio News Magazine of Kansas City, Missouri as associate editor. A monthly editorial appeared under his by-line. In addition, he created technical radio articles complete with graphics.

His written and photographic efforts have appeared in *Model Railroader* magazine, The Antique Collector, and other publications. He has produced articles for syndicated newspapers, compiled book reviews and was a consulting writer for the book *Norfolk & Western Railway, Volume 1*. His first book, *Pittsburgh & West Virginia Railway*, was published in 1989. Two other book manuscripts are being developed.

Current freelance writing assignments list historical, antiques and collectibles, life experience features. Howard's writing credits also include radio programming. He is a contributor to the Pittsburgh History Series of Pittsburgh public television station WQED.

Howard, his wife Dolly, and their Scottish Terrier "Major", reside in rural Butler County, Pennsylvania, near the historic village of Saxonburg. ✭